AF293479

Romina Amberg

Influence of *physical cues* from the degrading magnesium implants on human cells

Amberg, Romina: Influence of *physical cues* from the degrading magnesium implants on human cells, Hamburg, disserta Verlag, 2024

Buch-ISBN: 978-3-95935-614-5
PDF-eBook-ISBN: 978-3-95935-615-2
Druck/Herstellung: disserta Verlag, Hamburg, 2024
Covermotiv: pixabay.com

Charité – Universitätsmedizin Berlin
Charitéplatz 1, 10117 Berlin

Bibliografische Information der Deutschen Nationalbibliothek:
Die Deutsche Nationalbibliothek verzeichnet diese Publikation in der Deutschen Nationalbibliografie; detaillierte bibliografische Daten sind im Internet über http://dnb.d-nb.de abrufbar.

© disserta Verlag, Imprint der Bedey & Thoms Media GmbH
Hermannstal 119k, 22119 Hamburg
http://www.disserta-verlag.de, Hamburg 2024
Printed in Germany

Table of Contents

Abstract

Magnesium, as a biodegradable and biocompatible metal implant material for the orthopedic and vascular application, is also very promising in guided bone regeneration (GBR) to treat periodontal defects in dentistry. Therefore, thin magnesium membranes could replace the currently used barrier membranes made of resorbable collagen or non-resorbable titanium-reinforced polytetrafluorethylene (PTFE) for bone augmentation. Non-resorbable membranes are often reported to lead to exposed membranes as a result of postoperative wound dehiscence, and are accompanied by an increased infection risk and an extended healing period. Migration of human gingival fibroblasts (HGF) on the barrier membranes can overcome the exposed membrane by forming a cellular monolayer, which allows endothelial cells to finally close the wound. A migration assay has been developed during my previous research to investigate migration behaviour *in vitro* on magnesium surfaces. This study revealed that migration of HGF was slower on magnesium than on titanium and tissue culture plastic as control surfaces. During migration on magnesium, the cells were exposed to the alteration of various parameters, so called 'physical cues', involving surface alterations due to the formed corrosion layer and medium alterations arising from the dissolved corrosion products. Surface analysis by atomic force microscopy (AFM), scanning electron microscopy (SEM) and wettability measurements of all used materials revealed that the surface can be excluded as a parameter affecting migration rate on magnesium. Therefore, the main part of this study focused on analyzing the effect of the physical cues arising from the medium, like increasing Mg^{2+}, H_2 and osmolality as well as decreasing Ca^{2+} on HGF, including migration, viability and proliferation studies. It was shown that the altered ratio of Mg^{2+} and Ca^{2+}, caused by increasing Mg^{2+} and decreasing Ca^{2+} concentration, accompanied by an increase of H_2 concentration resulting from magnesium corrosion, led to reduced migration rate of HGF on magnesium surfaces. Furthermore, the individual increase of Mg^{2+} concentration up to 25 mM did not affect migration behaviour of HGF, while the migration rate decreased with decreasing Ca^{2+} concentration from 1.16 to 0 mM. This study provides detailed results of physical cues from the degrading magnesium membranes affecting migration, proliferation and viability of HGF. This knowledge allows us to create optimal conditions for the clinical application of magnesium membranes to ensure optimal healing success.

Zusammenfassung

Magnesium, als ein metallisches Imlantatmaterial für die orthopädische und vaskuläre Anwendung eignet sich aufgrund seiner Biodegradierbarkeit und Biokompatibilität hervorragend für die Anwendung in der gesteuerten Knochenregeneration (eng. Guided Bone Regeneration, GBR) zur Behandlung von parodontalen Defekten in der Zahnheilkunde. In Form von dünnen Membranen kann Magnesium die derzeit für den Knochenaufbau eingesetzten Barrieremembranen, bestehend aus resorbierbaren Kollagen oder nicht-resorbierbaren Titan verstärkten Polytetrafluorethylen (PTFE), ersetzen. Nicht-resorbierbare Barrieremembranen haben den Nachteil, dass häufig postoperative Wunddehiszenzen auftreten, währenddessen die Membran freigelegt wird, wodurch das Infektionsrisiko steigt und die Heilungsphase verzögert wird. Durch Migration von humanen gingivalen Fibroblasten (HGF) auf der Membran bildet sich ein zellulärer Monolayer, auf dem letztendlich die Endothelzellen migrieren können, um die offene Wunde zu schließen. Im Rahmen meiner bisherigen Forschungsarbeit habe ich einen Migrationsassay entwickelt, mit dem sich das Migrationsverhalten von Zellen *in vitro* auf Magnesiumoberflächen untersuchen lässt. Die Studie zeigt, dass die HGF auf der Magnesiumoberfläche langsamer migrieren, als auf der Titan- und der Kunststoffoberfläche, es aber dennoch zu einem Wundverschluss kommt. Infolge der Korrosion von Magnesium ändern sich die Umgebungsparameter, denen die Zellen während der Migration auf Magnesium ausgesetzt sind. Das sind zum einen Veränderungen im Medium durch gelöste Korrosionsprodukte und zum anderen Veränderungen der Magnesiumoberfläche aufgrund der Bildung einer Korrosionsschicht. Parameter, mit denen sich diese Veränderungen charakterisieren lassen, werden als ‚physical cues' bezeichnet.

Oberflächenanalysen mittels Rasterkraftmikroskopie, Rasterelektronenmikroskopie und Messungen zur Bestimmung der Benetzbarkeit zeigten, dass die Oberfläche keinen Einfluss auf die Migrationsrate der HGF auf den Magnesiummembranen hat. Demzufolge wird angenommen, dass nur die physical cues, mit denen sich das Medium charakterisieren lässt einen Einfluss auf das Zellverhalten haben. Auf dieser Annahme basierend, liegt der Fokus der Arbeit, insbesondere auf der Untersuchung der Auswirkungen einer gesteigerten Mg^{2+} Konzentration, H_2 Konzentration und Osmolalität sowie abnehmender Ca^{2+} Konzentration auf die Migration, Viabilität und Proliferation der HGF. Die Ergebnisse zeigten, dass sich die verlangsamte Migration der HGF auf Magnesium mittels der korrosionsbedingten Veränderung des Verhältnisses von Mg^{2+} und Ca^{2+}, sowie der gesteigerten H_2 Konzentration im Medium erklären lässt. Darüber hinaus hat der alleinige Anstieg von Mg^{2+} im Medium auf bis zu 25 mM keinen Einfluss auf die Migrationsrate der HGF, während die alleinige Abnahme der Ca^{2+} Konzentration von 1,16 zu 0 mM eine verlangsamte Migration zur Folge hat. Die Studie liefert detaillierte Ergebnisse über die Auswirkungen der physical cues der Magnesiummembranen auf das Zellverhalten der HGF hinsichtlich Migration, Viabilität und Proliferation. Mit diesem Wissen lassen sich optimale Bedingungen für die klinische Anwendung der Magnesium-Membranen schaffen, um einen optimalen Heilungserfolg anzustreben.

1 Introduction

Since the last century, magnesium-based implants have been probed in many studies for orthopedic and cardiovascular applications [1]. In contrast to other metal implants, magnesium is completely biodegradable by forming magnesium ions, molecular hydrogen and hydroxide ions in physiological conditions; hence, there is no need for a second surgery for implant removal [1]. Another advantage of magnesium is its biocompatibility due to magnesium ions being the 4th abundant cations in the human body and being involved as co-factors in numerous physiological reactions [1, 2].

Beside their biocompatibility and biodegradability, magnesium shows low elastic moduli similar to natural bone, thus preventing stress shielding of the regenerating bone [3]. Stress shielding occurs when the implant bears the mechanical load and relieves the healing bone, leading to delayed bone regeneration and in the worst case resulting in bone resorption [3].

Due to its excellent properties, magnesium could also be used in form of thin membranes for the guided bone regeneration (GBR) in dentistry to treat periodontal bone defects for inserting implant abutments [4]. As a first step of the GBR treatment, the defective bone is filled with bone substitute and the dental implant is inserted [5]. As a next step, the regenerating bone is covered by a barrier membrane, which is dressed by gingival tissue [5]. The GBR membrane provides a space for bone substitute replacement by natural bone and prevents ingrowth of the adjacent gingival tissue for optimal healing success [5].

Due to their excellent biodegradability and biocompatibility, currently used GBR-membranes are often made of collagen when simple bone augmentation is required [5]. In case of bone defects including alveolar wall defects, a barrier membrane with high volume stability is required. A current solution is non-resorbable titan-reinforced PTFE membranes, which provide high volume-stability due to its excellent mechanical stiffness [5]. However, the missing biodegradability requires a second surgery for implant removal after bone regeneration is completed, which is associated with a high burden for the patient, such as an extended healing period [5].

In this context, magnesium-based membranes combine adequate mechanical strength with the advantage of resorbability, which make magnesium a very promising implant material for GBR-membranes. Moreover, the reported antibacterial properties and the osteoproliferative effects of magnesium, resulting from the release of magnesium ions and hydroxide ions during magnesium corrosion, provide a great advantage compared to the current used membranes [6, 7]. In regard to non-resorbable GBR-membranes based on titanium, membrane exposure (Fig.1) provoked by insufficient tissue coverage for primary wound closure or wound dehiscence during healing is often reported as the main problem of non-resorbable membranes due to the high infection risk [8]. Wound dehiscence arise in 31% of all GBR procedures using non-resorbable GBR-membranes, resulting in early membrane removal associated with an decreased regeneration rate (0-60%) [9]. An exposed membrane can be overcome by migration of human gingival fibroblasts (HGF) on the GBR membrane to form a cellular monolayer, which allows epithelial cells to migrate to the top of the HGF to finally close the wound [10].

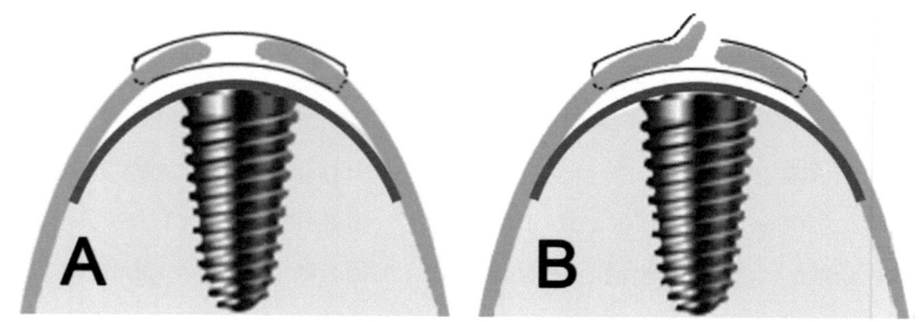

Figure 1. Membrane exposure due to incomplete soft tissue coverages after surgery (A) or due to wound dehiscence (B) [4].

This clearly demonstrates the high relevance of investigating the migration behaviour of HGFs on GBR-membranes, especially on magnesium as a novel membrane material. During my previous research, I developed a scratch-based migration assay on magnesium membranes, including fluorescent cell staining and pre-corrosion of magnesium, which allows studying wound closure of HGF *in vitro* on biodegradable magnesium under life view conditions [4]. The results showed that migration is slower on magnesium surfaces than on tissue culture plastic surfaces. To obtain more reliable results, current used membrane materials such as titanium surfaces should be tested with the established migration assay to compare the migration behaviour. Furthermore, during cell migration on magnesium membranes, the HGF are affected by the medium alterations arising from the dissolved corrosion products and by surface alteration due to the formation of a corrosion layer. All parameters characterizing the environmental parameter and the surface properties of a material are so called 'physical cues'. Each physical cue affects cell proliferation, viability and migration of HGF and needs to be considered in the clinical application of magnesium membranes. In regard to medium alteration, the magnesium ions concentration, osmolality, pH value and molecular hydrogen concentration increase [11], and calcium ion concentration decreases [12] due to magnesium corrosion. The surface condition can be characterized by wettability, roughness and topography measurements. Therefore, the current study included surface characterization of magnesium, as well as titanium and tissue culture plastic as control surfaces by using atomic force microscopy, electron scanning micro-scopy and wettability measurements.

The investigation of physical cues from degrading magnesium affecting HGF, especially cell migration, offers new opportunities for the clinical application of magnesium membranes for the GBR and is needed for ensuring optimal healing success. Moreover, cell migration is essential in any regeneration processes and particularly on implanted device surfaces, emphasizing the high relevance for migration studies on magnesium membranes as a novel implant material.

Hence, the aim of this study was to analyze migration behaviour of HGF on magnesium membranes by investigating the physical cues from the degrading magnesium and their separate effect on cell migration, proliferation and viability.

2 Methods

2.1 Cell culture

All experiments were performed with human gingival fibroblasts (HGF-1 CRL-2014, ATCC, Manassas, USA), which were cultivated in cell culture medium (CCM) under cell culture conditions (37 °C, 20% O_2, 5% CO_2, 95% rH) in a Hera-cell240 incubator (Heraeus, Hanau, Germany). The CCM consists of phenol red free Dulbecco's modified Eagle medium (DMEM)/F-12, penicillin/streptomycin (100 U ml-1, 100 µg ml-1) (both Life Technologies GmbH, Karlsruhe, Germany) and 10% fetal bovine serum superior (FBS, Biochrom GmbH, Berlin, Germany). At 80-90% confluence under the DMIL microscope (Leica, Wetzlar, Germany), the cells were splitted in new cell culture flasks (Sarstedt Nürnbrecht, Germany) by rinsing the cells with phosphate buffered saline solution (PBS, Life Technologies GmbH), and were then detached with trypsin (0.25% Trypsin-EDTA solution, Sigma-Aldrich, Steinheim, Germany) for 2 min. Cell refeeding with fresh CCM ensued every 2-3 days.

2.1 Migration assay

To compare the migration behaviour of human gingival fibroblasts (HGF) on magnesium membranes (Botiss biomaterials GmbH, Berlin, Germany, Purity: 99.95%, 13 x10 mm, thickness: 140 µm) with them on current used non-resorbable GBR-membranes, the established migration assay, which is described in my paper [4] was performed on gamma-sterilized titanium discs (Goodfellow, Friedberg, Germany, Purity: 99.6%, Ø15 mm, thickness: 140 µm). The migration curve, obtained from the evaluated images using the software Tscratch, represents the averaged cell free area of six replica as a function of the time.

2.2 Surface characterization

The roughness and topography of uncorroded and 72 h pre-corroded gamma-sterilized magnesium membranes, which were used for the migration assay, were already determined during my previous research [4]. The analysis was completed by roughness measurements with the atomic force microscopy (AFM) of titanium discs and tissue culture plastic (TCP, 24-well plate, Sarstedt, Nürnbrecht, Germany) and topography images of titanium discs using the scanning electron microscopy (SEM). Additionally, the wettability was determined for magnesium (uncorroded and pre-corroded), titanium discs and TCP. The utilized surface methods are described in the following sections.

2.2.1 Topography - Scanning electron microscopy (SEM)

In contrast to light microscope, the scanning electron microscope (SEM) uses electrons for imaging, whereby a magnification of up to 10^5 x and a resolution of 1-5 nm can be achieved [13]. The high depth of sharpness and the high magnifications of the SEM allow capturing of high-resolution topography images. Samples need to be electrically conductive and resistant to vacuum. Non-conductive sample can be made electrically conductive by coating with a thin gold layer in a sputtering process [14]. But, there are already modern SEM working with low vacuum to investigate non-conductive samples without sputtering [15].

An electron gun generates a beam of electrons by thermal emission (Fig.2) [14]. The electrons dissolve from the cathode and were accelerated to the anode under vacuum. The acceleration voltage determines the energy of the electrons leaving the anode, which is between 2-30 keV [14]. Operating under vacuum avoids electrons to collide with gas molecules and allows using the wave properties of electrons [16]. Electrons were focused by the condenser lens to a small electron beam, which is scanned in a raster pattern over the sample surface [14]. Scanning is achieved by the beam deflector containing scanning coils. When electrons hit the sample, they interact elastically or inelastically with the sample atoms [17]. During elastically interaction, the electron is scattered by the positive atom nucleus. The generated backscattered electrons (BSE) are characterized by a low energy loss and were detected with a semiconductor detector [17]. Inelastically interaction occurs when the impinging electron knocks another electron from its atomic shell, which is called secondary electron (SE) [17]. Secondary electrons (SE) show a high energy loss with a high deflection angle and are detected with the Everhart-Thornley detector, a type of scintillation-photomultiplier system [17]. The detector signal is amplified and converted in an indirect image on the screen. Every voltage signal is assigned to a brightness value. Topography images are principally captured using SE-signals, whereas BSE signals provide information about the chemical composition of the sample [14]. Although SE are generated in the whole sample, the generated SE in deeper layers cannot escape the sample due to their low energy and are thus absorbed. Therefore, only the SE in direct vicinity to the interface form the signal (escape depth for metals: 5 nm and for isolators: 50 nm), which are 1% of all formed SE [18].

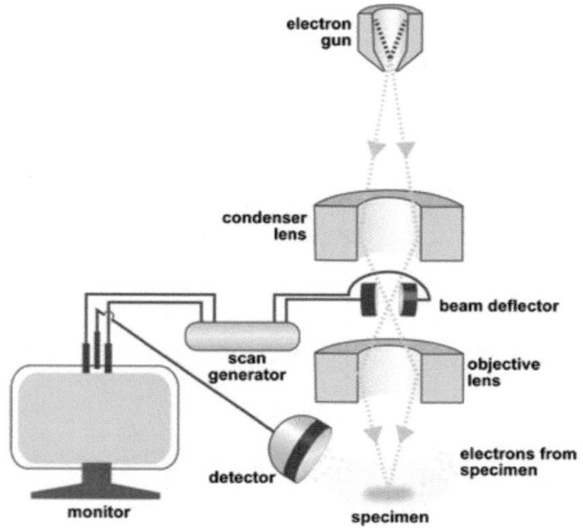

Figure 2. The basic components of the scanning electron microscope (SEM) [17].

Nontheless, these SE signals create high-resolution images of the topography of the sample surface. An elevated point of the surface leads to forming a higher SE amount, resulting in a higher voltage signal [18]. The higher the voltage signal, the brighter the pixel captured in the image [18]. The titanium discs were measured with the SEM JCM-6000Plus NeoScope™ (Joel, Tokyo, Japan) operating with the high vacuum mode at an acceleration voltage of 5 kV using the SE signals to image the topography of the titanium surface.

2.2.2 Roughness - Atomic force microscopy (AFM)

The atomic force microscopy (AFM) enables roughness measurements of micro- and nanostructures by imaging topographical profiles [19]. The mean roughness R_a is the most common surface roughness parameter, and is the arithmetic averaged height of roughness-component irregularities from the mean line, measured within the sampling length l_r (Fig.3) [20].

$$R_a = \frac{1}{l_r} \int_0^{l_r} |z(x)|\, dx \qquad\qquad (1)$$

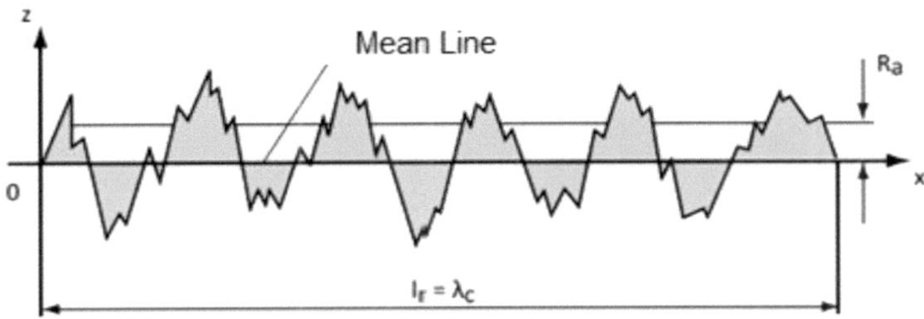

Figure 3. Mean roughness R_a for a roughness profile [21].

Further 2D parameters are the mean roughness depth R_z and the root mean square R_q [22], which are all obtained from a single line of the surface. The 3D parameters S_a, S_q and S_z, are used for characterization of defined area, and are the averaged value of all line profiles [22].

The measuring principle is based on scanning a surface with a very sharp tip, which is fixed on a cantilever using the interactions between a tip and a sample surface to get a topographical image of the sample surface [23]. The sample is moved in the x, y and z directions by a piezoelectric material, thus position can be controlled in nanometer resolution [24]. The tip is generally made of silicon (Si), silicon nitride (Si_3N_4) or silicon dioxide (SiO_2) with a contact radius of several nm up to 100 nm [25]. A laser beam is focused onto the back of the cantilever and is reflected back to a photodiode detector (Fig.4). Depending on the atomic force variations between the tip and the sample surface, the deflection of the cantilever can be measured precisely and detected as a topographical image.

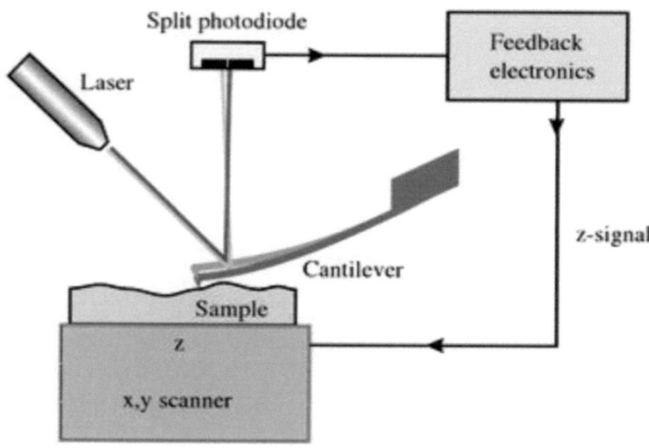

Figure 4. Schematics of atomic force microscopy (AFM) operation [24].

There are two operation modes: the contact mode and the non-contact mode depending on whether the tip touches the surface during the measurement. In the contact mode, the signal can be set to a constant height (constant height mode) or a constant force (constant force mode) of the cantilever [23]. When operating in constant height mode, the tip scan the surface, while the cantilever bends and supplies a signal to the detector [23, 24]. Maintenance of constant force between the tip and the cantilever (constant force mode) is achieved using a control loop to obtain a constant photodiode signal [23, 24]. The feedback signal of the control loop contains information of the surface topography [23]. The contact mode has a high resolution and a fast measurement time; the disadvantages are that the lateral resolution is limited due to the diameter of the tip and that the sample can be destroyed, thus modified results may be obtained [24].

The disadvantages of the contact mode were the reason for developing the non-contact mode in 1987 by Martin *et al.* [23]. In the non-contact mode the cantilever oscillates at its resonance frequency in a distance of 10-100 nm from the surface [23, 25]. Variations in the distance between the surface and the tip lead to alter the resonance frequency or vibration amplitude [19, 26]. On the one hand, the frequency alteration can be measured at constant amplitude (FM-AFM); on the other hand, the amplitude alteration can be measured at constant frequency (IC-AFM) [26]. Therefore, the respective measurement variable feeds the control loop.

The titanium and tissue culture surfaces were measured in the non-contact mode (IC-AFM) with the atomic force microscope easyScan 2 (Nanosurf®, Liestal, Switzerland).

The tip was made of silicon and had a contact diameter of 7 nm (Nanosensors Typ PPP-NCLR-10, Switzerland). The 3D mean roughness values S_a were determined using the software Gwyddion 2.45.

2.2.3 Wettability measurements

The wettability describes the relative adhesion of a liquid such as water or a solvent to a solid surface due to intermolecular interactions [27]. The contact angle measurement is a common method used to determine the wettability of a surface. Therefore, the contact angle θ between the sessile drop and the sample surface is measured. Contact angles less than 90° reveal wettable surfaces, while contact angles above 90° were obtained for non-wettable surfaces (Fig.5). In cases where water is used to wet the surface, hydrophilic surfaces have contact angles less than 90° and hydrophobic surfaces above 90°.

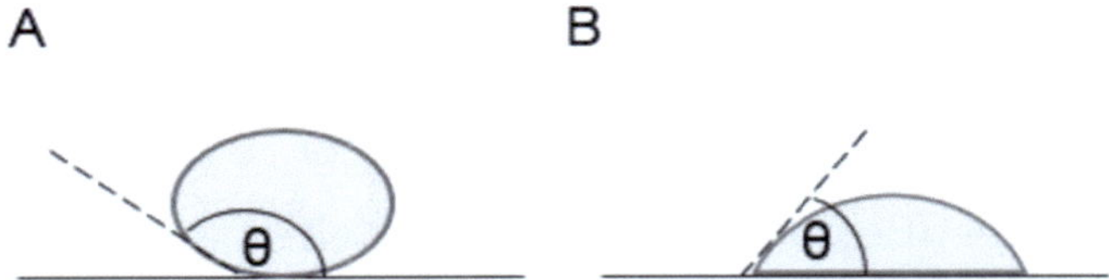

Figure 5. Contact angle above 90° (A) and below 90° (B).

The surface free energy (SFE, σ_s) is the surface tension of a solid and can be determined by measuring the contact angle of more than one liquid using the model of Owens, Wendt, Rabel and Kaelble (OWRK-model) [28, 29], which is based on the Young equation.

The Young equation describes the relationship between the surface free energy σ_s of the sample surface, the interfacial tension σ_{sl} between liquid and solid, the surface tension σ_l of the liquid and the contact angle θ [27]:

$$\sigma_s = \sigma_{sl} + \sigma_l \cdot \cos\theta \tag{2}$$

In contrast to the wettability, described by a contact angle of one liquid, the SFE is independent of the liquid used. One can predict the behaviour of any liquid on the surface by knowing the SFE.

In our case, the contact angles of water and diiodomethane were measured on titanium, magnesium (pre-corroded, uncorroded) and tissue culture plastic (TCP) surfaces using the Drop Shape Analyzer DSA 25 (Krüss GmbH, Nürnberg, Germany) with the sessile drop method. The SFE were calculated from the measured water contact angle (θ_{water}, WCA) and diiodomethane contact angle ($\theta_{diiodom.}$, DCA) and the surface tension of the liquids (σ_{water} = 72.8 mN m^{-1}, $\sigma_{diiodom.}$ = 50.8 mN m^{-1}) using the equation (2).

2.3 Modified migration assay

The established migration assay allows migration studies directly on non-transparent surfaces. To verify whether surface alterations of magnesium surface affects cell migration, the migration assay was modified. To exclude the direct surface contact of the cells to the pre-corroded magnesium membrane, the HGF were seeded on the bottom of the 24-well plate and affected by the corroding magnesium membranes, at a distance of 2 mm from the cells. In accordance with the established migration assay, the utilized magnesium membrane was pre-corroded for 72 h in 10 mL CCM in a tightly closed 15 mL falcon tube (Becton & Dickinson, Heidelberg, Germany) on a roll mixer (Phoenix, Garbsen, Germany) under cell culture conditions. After labeling the HGF with 10 mM fluorescent dye CellTracker^{TM} Red CMTPX (Life Technologies GmbH) for 30 min using serum free CCM, the cells were cultured for 24 h in CCM, before cell seeding (4·10^4 cell/ 1 mL/ well) into the silicone inserts ($\emptyset_{inner}$: 10 mm) of the 24-well plate. The silicone inserts fit precisely into the wells of the plate and were made of transparent autoclavable casting compound of silicon elastomer (Sylgard® 184, Sigma-Aldrich, Steinheim, Germany). Cells were allowed to adhere for 24 h. To ensure conditions comparable to the established migration assay, the pre-corroded magnesium membrane was also incubated with 1 mL CCM for 24 h. After removing the silicone inserts, the confluent cell monolayer was scratched with a sterile pipette tip (10 µL), rinsed with serum free CCM, and the pre-corroded magnesium membrane was placed on a silicone half ring identical to the original assay with the only difference being that the cells do not grow on the magnesium membrane but on the plastic bottom.

The silicone half rings ($\emptyset_{inner}$: 10 mm, $\emptyset_{outer}$: 14 mm, thickness: 2 mm) were generated by stamping the autoclavable silicone sheet (Exact Plastics GmbH, Bröckel, Germany) and creating four tunnels to ensure optimal medium exchange. The set-up was completed by adding 2 mL CCM to the cells and fixing the membrane with two glass stones, respectively. The plate was incubated for 4 h under cell culture conditions and then tipped in a 45° angle before the 72-h imaging period began. This step was crucial to remove the newly formed hydrogen bubbles due to highly initial corrosion rate of magnesium, which would impair the imaging. The images were taken every 5 h using the DMI6000 B inverted fluorescent microscope (Leica, Mannheim, Germany) with a live cell imaging system and evaluated with the software TScratch (see 2.1).

The cell migration behaviour on magnesium (Mg-direct) from the previous experiments [4] were compared to them, which were only affected by the corroding medium (Mg-indirect).

2.4 Characterization of the supernatants during cell migration

Both migration assays (Mg-direct, Mg-indirect) were performed and the hydrogen concentration of the supernatant was measured with the Hydrogen Microsensor Multimeter (Unisense, Aarhus, Denmark) at starting point (0 h), after 25 h and after 50 h, by measuring each sample for 30 s without stirring. The supernatants were removed and stored at -20 °C for several days until shipping on dried ice to Freiberg University of Mining and Technology (Germany) for further investigations. The quantification of Mg^{2+} and Ca^{2+} was determined by inductively coupled plasma optical emission spectrometry (ICP-OES) using an iCAP 6000 from Thermo Fisher Scientific (Bremen, Germany) with an Echelle grating optical system and a Charge Injection Device (CID), type CID86 detector. Nebulizer was PEEK (Polyetheretherketon) MiraMist (Burgener, Mississauga, Canada). Method parameters were the following: Rf power 1100 W; radial viewing plasma mode; coolant gas flow 12.0 L min^{-1}; auxiliary gas flow 0.6 L min^{-1}; nebulizer gas flow 0.65 L min^{-1}. Sample digestions were prepared with 2% nitric acid made of ultrapure deionized water with a conductivity of 0.055 µS/cm, obtained from MicroPure water purification system from TKA (Niederelbert, Germany), and nitric acid from Merck KgaA (Darmstadt, Germany) that was bidistilled in-house with PFA sub-boiling acid distillation system DST-100 from AHF Analysentechnik AG (Tübingen, Germany). For quantification, external calibration models with dilutions of stock solutions from Merck KgaA were used. Emission lines 279.55 nm, 280.27 nm and 285.21 nm for magnesium and emission line 396.85 nm for calcium were measured and evaluated.

The osmolality and pH value of the supernatants after migration on magnesium membranes (Mg-direct) were determined by previous research [4].

2.5 Ca^{2+} determination in the corrosion layer of Mg membranes

While performing the unmodified migration assay (Mg-direct), the magnesium membranes were removed at starting point (0 h), after 25 h and after 50 h and stored at -20 °C for several days until shipping on dried ice to Freiberg University of Mining and Technology, for Ca^{2+} determination. After digestion of each Mg membrane with 1 mL nitric acid (69%), the samples were diluted with ultrapure deionized water and measured with ICP-OES.

2.6 Immunohistological analysis

The effect of increasing Mg^{2+} in the medium on cell adhesion of HGF was investigated by immunohistological staining. Therefore, the HGF were seeded in a 48-well plate (4·10^4 cells/well) and allowed to adhere for 24 h. Then, the cells were incubated with 25 mM and 75 mM MgCl$_2$ · 6 H$_2$O for 24 h, rinsed with serum free CCM, and fixed with polymeric formaldehyde (PFA) for 10 min (400 µL/well). Cells were rinsed three times for 3 min with TRIS-buffered solution (TBS) washing buffer (pH 8.2, 0.025% Triton) and permeabilized with 0.25% Triton in TBS (pH 8.2) for 10 min. After washing with TBS washing buffer, the cells were treated with blocking solution for 30 min to block unspecific bindings to the substrate. The blocking solution contained 5% goat serum and 1% BSA (bovine serum albumin) in TBS (pH 8.2, without Triton) (both Abcam, Cambridge, UK). Then, the blocking solution was removed and the cells were incubated with the Anti-Vinculin mouse antibody (Merck, Darmstadt, Germany) for 17 h at 4 °C, which was diluted 1:150 with antibody diluent (Merck) (150 µL/well). The antibody solution was removed and the cells were rinsed three times for 5 min with TBS washing buffer. Cells were incubated with a mixture of the goat Anti-mouse IgG Alexa Fluor® 488 secondary antibody (1:400 dilution) and Alexa Fluor® 594 Phalloidin (1:100 dilution) (both Thermo fisher scientific Molecular Probes, Waltham, USA) in TBS (pH 8.2, without Triton) for 1 h, containing additionally 1% BSA and 5% goat serum

(150 µL/well). The incubation was performed in the dark. After rinsing the cells three times for 5 min with TBS washing buffer and once with sterile injection water (B Braun, Penang, Malaysia), the cell nucleus was stained with 4',6-Diamidin-2-phenylindol (DAPI, 1:1500 dilution with sterile injection water) by incubating for 15 min. After removing DAPI, the well plate was stored at 4 °C, containing TBS washing buffer until imaging was performed with the Axio Observer fluorescence microscope (Zeiss, Jena, Germany).

2.7 Preparing test solutions

To investigate the effect of increasing Mg^{2+} and osmolality, as well as decreasing Ca^{2+} concentration on cell behaviour, various saline solutions including controls were prepared. Increased hydrogen concentration was implemented by enriching cell culture medium with molecular hydrogen. The effect of pH rising was not separately tested, as previous experiments showed that pH does not increase significantly during migration on magnesium implants due to the buffered system [4]. The combined effect of all altered physical cues during magnesium corrosion on cells was tested by preparing various magnesium extracts.

<u>Saline Solutions</u>

Cell culture medium (CCM) was supplemented with magnesium chloride hexahydrate ($MgCl_2 \cdot 6\ H_2O$), mannitol and sodium chloride (NaCl), respectively. The concentrations are declared in Table 1. Mannitol and NaCl were used as controls, as mannitol matches the osmolality and NaCl the chloride concentration, which have to be considered by adding $MgCl_2 \cdot 6\ H_2O$ to the CCM. Additionally, test solutions with different Mg^{2+} concentrations were prepared, which all show same ionic strength and osmolality, achieved by adding various amounts of mannitol and NaCl to the magnesium test solutions. The effect of lower calcium ion concentration was investigated using calcium and phenol free Dulbecco's modified Eagle medium (DMEM, United States Biological, Salem, USA), which was supplemented with 10% FBS and calcium chloride dihydrate ($CaCl_2 \cdot 2\ H_2O$) (Tab.1). All utilized substances were provided by Merck (Darmstadt, Germany), and the prepared solutions were sterile filtrated with the Millex-GV (0.22 µm).

Table 1. Analysis of tested saline solutions and magnesium extracts [4].

		Mg^{2+} [mM]	Cl^- [mM]	Ionic strength [mM]	Osmolality [mOsmol Kg^{-1}]	Ca^{2+} [mM]
control	CCM*	0.7	124	---	315±6	1.16
saline solution**	25mM MgCl$_2$	25	50	75	369±6	---
	75mM MgCl$_2$	75	150	225	480±5	---
	Mannitol-1	0	0	0	369±6	---
	Mannitol-2	0	0	0	480±5	---
	50mM NaCl	0	50	50	406±2	---
	150mM NaCl	0	150	150	580±8	---
	0 mM Mg^{2+}	0	75	75	450±3	---
	15 mM Mg^{2+}	15	60	75	450±3	---
	25 mM Mg^{2+}	25	50	75	450±3	---
	0 mM Ca^{2+}	0.7	0	---	307±1	0
	0.3 mM Ca^{2+}	0.7	0	---	308±1	0.3
	0.5 mM Ca^{2+}	0.7	0	---	309±1	0.5
	0.8 mM Ca^{2+}	0.7	0	---	310±1	0.8
Mg-extract	w/o CO$_2$	4.5±0.1	---	---	330±3	0.93±0.02
	w CO$_2$	58.2±0.8	---	---	385±3	0.57±0.01

*Cell culture medium

**Values of Mg^{2+}, Ca^{2+}, Cl^- and ionic strength are calculated values and do not include the absolute concentration of the cell culture medium.

<u>Mg-extracts</u>

Mg-extracts were obtained by immersion a magnesium membrane in 10 mL CCM in a 15 mL falcon tube and leaving on a roll mixer for 72 h under cell culture conditions. To obtain two different Mg-extracts, one tube was tightly closed to avoid environmental exchange and another tube was provided with a sterile filter, which was attached to the lid of the tube to ensure environmental exchange. The Mg^{2+} and Ca^{2+} concentrations were determined at Freiberg University of Mining and Technology with ICP-OES (Tab.1). The osmolality of all tested saline solutions and Mg-extracts were determined in triplicate with the osmometer (Gonotec, Berlin, Germany) and are indicated in Table 1.

Hydrogen-rich cell culture medium (H-CCM)

Cell culture medium (CCM) was enriched with molecular hydrogen using the Highdrogen Age$_2$ Go hydrogen generator (Aquacentrum, München, Germany). Hydrogen was generated for 3 min with open bottle, followed by 20 min with closed bottle, containing 125 mL CCM. The hydrogen-rich cell culture medium (H-CCM) was sterile filtrated and diluted with normal CCM at dilutions of 1:2 and 1:4 to obtain medium with three different H$_2$ concentrations.

2.8 Toxicological assays (MTT, BrdU) with test solutions

To investigate the anti-proliferative effect of the prepared test solutions (see 2.7) on human gingival fibroblasts (HGF) a 5-bromo-2'-deoxyuridine (BrdU) assay was performed. The effect on cell viability was investigated with the methyl-thiazolyl-tetrazolium (MTT) assay (both Roche diagnostics, Penzberg). The BrdU assay detects incorporated BrdU into DNA during cell proliferation after addition of substrate labelled antibody (Anti-BrdU), and the MTT assay measures the metabolic activity of the cells by converting the MTT substrate into formazan.

The HGF were seeded in a 96-well plate (4 000 cells/well) and were allowed to achere for 24 h before incubating with the test solutions for another 24 h. After removing the test solutions, the cells were rinsed with serum-free CCM and fresh medium (100 µL/well) was added. The rinsing step is required to remove magnesium ions, which would interact with the MTT reagent leading to false positive errors.

Then, the BrdU and MTT assay were performed according to the manufacturer's protocol. For the BrdU assay, 10 µL/well BrdU labeling solution was added to the cells (final concentration: 10 µM BrdU/well) and incubated for 4 h under cell culture conditions. After removing the BrdU labeling solution by tapping off, 200 µL/well FixDenat was added for 30 min to fix the cells. FixDenat was removed by tapping and 100 µL/well anti-BrdU-POD working solution was added for 90 min, and then removed by tapping. Cells were rinsed three times with 200 µL/well washing solution and 100 µL/well substrate solution was added. Absorption was measured after 15 min incubation at 370 nm and 492 nm using an ELISA-Reader type infinite M200pro (Tecan Group AG, Männedorf, Switzerland). For the MTT assay, 10 µL/well MTT labeling reagent was added and incubated for 4 h under cell culture conditions. Then,

100 µL/well solubilization solution was added and incubated for 18 h under cell culture condition. Absorption was measured at 555 and 670 nm using the ELISA-Reader. Due to the rapid clearing of hydrogen from the medium, in another well plate, the medium was additionally replaced by fresh H-CCM every 30 min for the first 2.5 h. Then, BrdU and MTT assays were performed. Hence, the cells were affected by a higher hydrogen concentration over a longer period.

2.9 Migration assay with test solutions

The migration assay was performed based on the modified migration assay (Mg-indirect) (see 2.3). Therefore, the labelled HGF were seeded in a 48-well plate to ensure equal cell growth area to the modified migration assay. After scratching and rinsing the confluent cell monolayer, the test solutions were added, and imaging was started. The hydrogen concentration of the various H-CCM and the Mg-extracts were measured up to 60 min in another well-plate, which was placed in the incubator.

2.10 Data and statistical analysis

The measured data were analyzed using descriptive statistics to clarify fundamental effects of the physical cues from the degrading magnesium implants on human cells. Therefore, the data were expressed as the mean ± standard deviation, and the graphics were plotted with Microsoft Excel® software (MS Excel 2016, Washington, USA). To compare the means of two groups, an independent samples t-test was calculated using the Statistical Package for the Social Science (SPSS, v22, Chigago, USA). P-values less than 0.05 were considered statistically significant.

3 Results

3.1 Migration behaviour on Mg, Ti and TCP

The established migration assay allows cell migration studies on pre-corroded magnesium surfaces [4]. To compare the migration behaviour of HGF on magnesium with them on current used membrane materials, the same assay was performed with titanium discs. The results showed that HGF migrate the slowest on pre-corroded magnesium, followed by tissue culture plastic (TCP, control) and the fastest on titanium surfaces (Fig.6). To compare the migration behaviour quantitatively, the value of 50% scratch-area filling of the initial cell-free area was determined for each material. On magnesium, the time $t_{50\%}$ is 22 h, on titanium the $t_{50\%}$ is 8 h and on control surface the $t_{50\%}$ is 13 h.

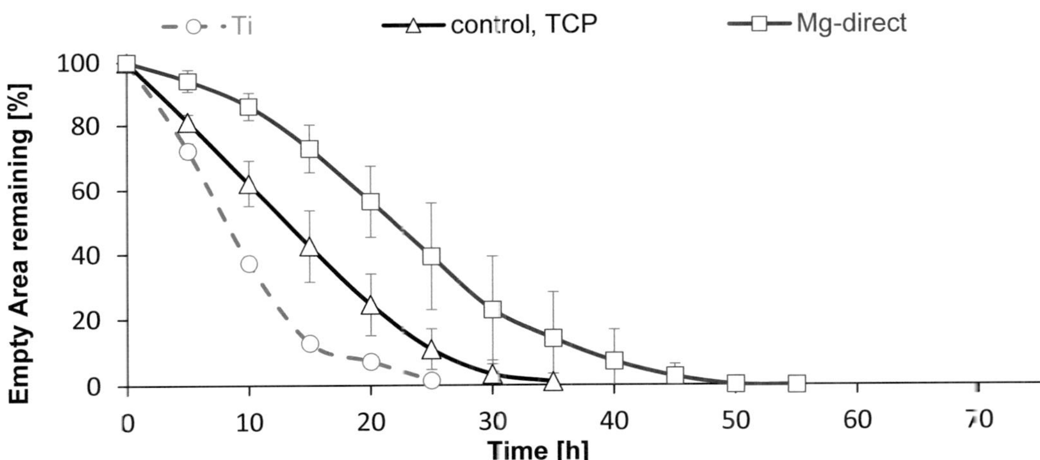

Figure 6. Process of wound closure *in vitro* of HGF on magnesium (Mg-direct), titanium (Ti) and tissue culture plastic (TCP, control) measured on an inverted microscope with a life cell imaging system utilizing the remaining empty scratched area to determine the process of cell migration over time. Data represent the mean±SD (n=4) [4], modified.

3.2 Characterized Mg, Ti and TCP surfaces

To assess the different migration behaviour of HGF on Mg, Ti and TCP, the surfaces of all materials were analyzed. The surface roughness and topography measurements for Mg that were performed during my previous research were completed for Ti and TCP. During corrosion of Mg, the surface becomes obvious with corrosion cracks, which is typical for Mg immersed in physiological solution or cell culture medium [30]. In contrast, the Ti surface exhibits smooth surface with additional parallel microgrooves (Fig.7).

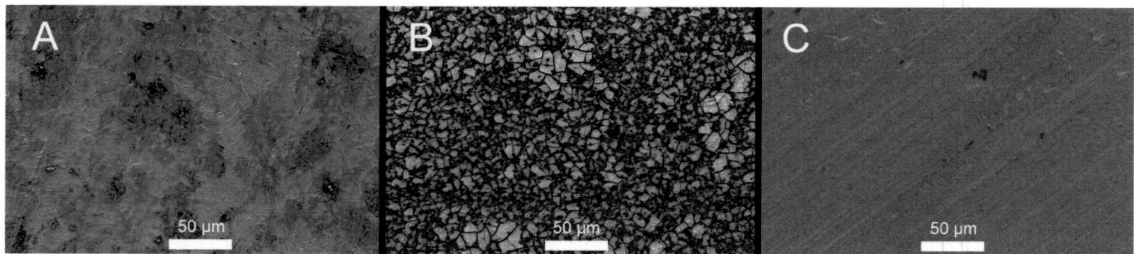

Figure 7. Topography of uncorroded (A), pre-corroded (B) Mg membrane and Ti disc (C) captured with the scanning electron microscopy (SEM), 400x magnification [4].

The roughness of the control surfaces TCP and Ti are significantly smoother compared to magnesium membranes, as the S_a values are much lower, indicated by S_a (Ti) = 74.5 ± 8.7 nm and S_a (TCP) = 10.7 ± 2.4 nm. Pre-corrosion of Mg did not significantly alter the roughness. Pre-corroded Mg has a S_a value of 131.6 ± 4.4 nm, while uncorroded has a S_a value of 129.7 ± 2.2 nm (Fig.8).

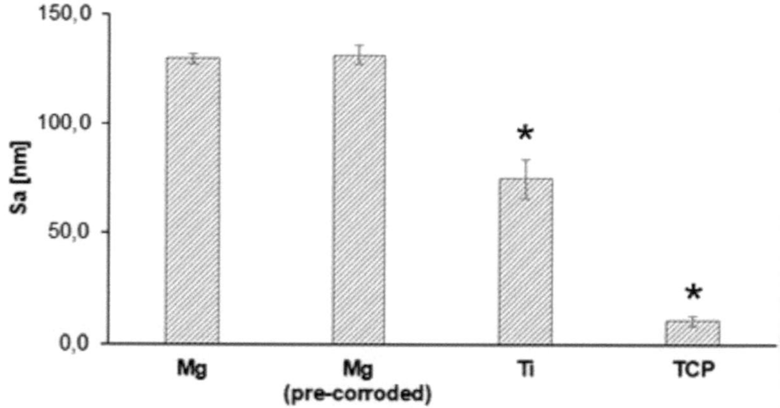

Figure 8. Arithmetic average 3D roughness values (S_a) utilizing atomic force microscopy (AFM) to investigate Mg membranes (uncorroded and pre-corroded), Ti and TCP. Data represent the mean±SD (n=4). Asterisk indicates significant difference (p<0.05) to both other test materials [4].

The results of the wettability measurements of Mg, Ti and TCP with the Drop Shape Analyzer are depicted in Figure 9. The water contact angle (WCA) of pre-corroded Mg was the highest (77.8 ± 5.4°), followed by Ti (72.7 ± 7.3°), uncorroded Mg (69.7 ± 2.6°), and the lowest for TCP (48.2 ± 7.9°). Lower WCA values indicate more hydrophilic surfaces, as the surface is readily wettable with water. The surface free energy (SFE) ensues from the WCA and diiodomethane contact angle (DCA). TCP shows the highest SFE (γ = 57.2 ± 7.1 mN m^{-1}), which is in line with the measured hydrophilic WCA value, as surfaces with higher SFE try to lower their energy by adsorbing low energy materials such as hydrocarbon [31].

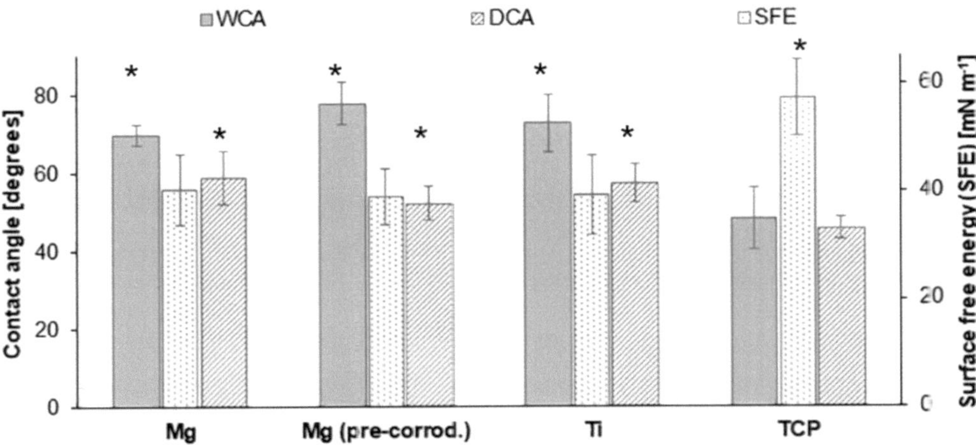

Figure 9. Contact angle for water and diiodomethane of Mg (uncorroded and pre-corroded), Ti and TCP and the resultant surface free energy (SFE). Data represent the mean±SD (n=5). Asterisks represent significant differences compared to the other test materials (p<0.05) [4].

3.3 Mg^{2+} affects cell adhesion

The effect of Mg^{2+} on HGF cell adhesion was investigated by immunohistological cell staining of vinculin and actin, including nucleus staining with DAPI after incubation of 24 h with cell culture medium (CCM) containing MgCl$_2$ · 6 H$_2$O (Fig.10). Cells, which were treated with 75 mM MgCl$_2$, showed more green-stained vinculin spots and a more spread cell body compared to cells, treated with 25 mM MgCl$_2$ and the control (treated with CCM). There was no visible difference between the 25 mM MgCl$_2$ and the control cells regarding cell morphology and vinculin amount.

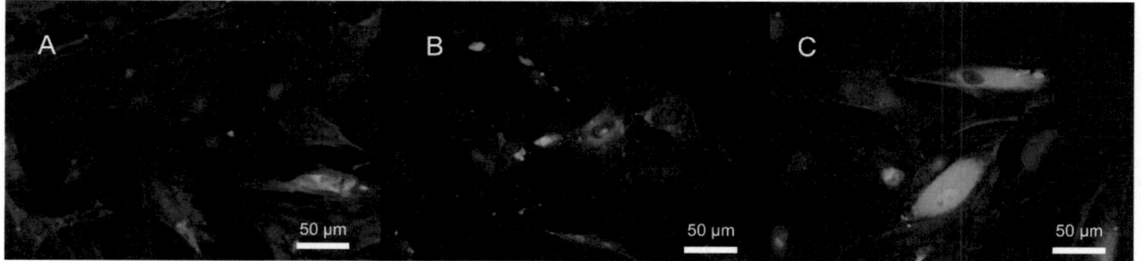

Figure 10. HGF in 80% confluent 48-well plate, treated with cell culture medium (A), containing additionally 25 mM (B) and 75 mM (C) MgCl$_2$ for 24 h and stained for vinculin (green), actin (red) and the nucleus (blue), 40x magnification [4].

3.4 Unmodified vs. modified migration assay (Mg-direct/ Mg-indirect)

The established migration assay (Mg-direct) was modified to exclude the surface effect of corroding magnesium on HGF, ensuring that the cells were only affected by the altered medium (Mg-indirect). The migration curves of both assay variants showed a comparable migration behaviour (Fig.11). With the modified migration assay (Mg-indirect), the time of 50% scratch-area filling of the initial cell-free area was 19 h, while the time was 22 h when HGF migrate directly on the magnesium surface (Mg-direct). The independent samples t-test was performed only for the 20 h time point to compare the migration behaviour between each migration assay (Mg-direct/ Mg-Indirect) and the control in the middle of the migration range.

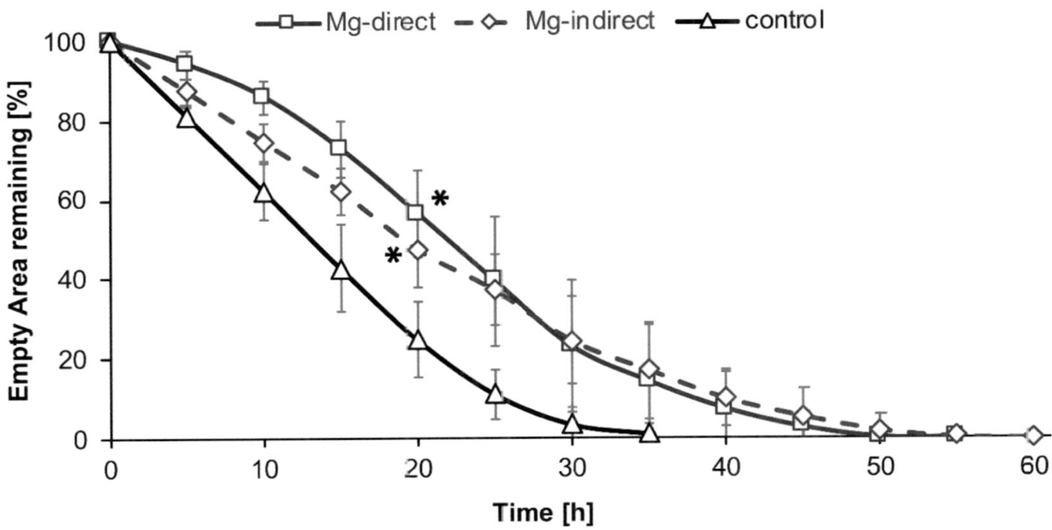

Figure 11. Process of wound closure *in vitro* of HGF on pre-corroded Mg membrane with surface contact (Mg-direct) and without surface contact (Mg-indirect), measured on an inverted microscope with a life cell imaging system utilizing the remaining empty scratched area to determine the process of cell migration over time, control: cell culture medium. Data represent the mean±SD (n=4). Asterisks represent significant differences compared to the control (p<0.05) [32].

3.5 Characterized supernatants during cell migration

To confirm the same medium conditions in both migration assays, the supernatants of the HGF were characterized for Mg^{2+}, Ca^{2+} and H_2 concentrations over time. The data for the unmodified (Mg-direct) and modified (Mg-indirect) migration assay, shown in Figure 12, reveal similar trends among each other. The Mg^{2+} concentration increased slightly to 8.5±1.9 mM and the Ca^{2+} concentration decreased slightly over time from 1.17±0.07 mM to 0.96±0.11 mM. The H_2 concentration was significantly higher than that of the control, indicated by approximately 180±0.20 mM over the entire migration time.

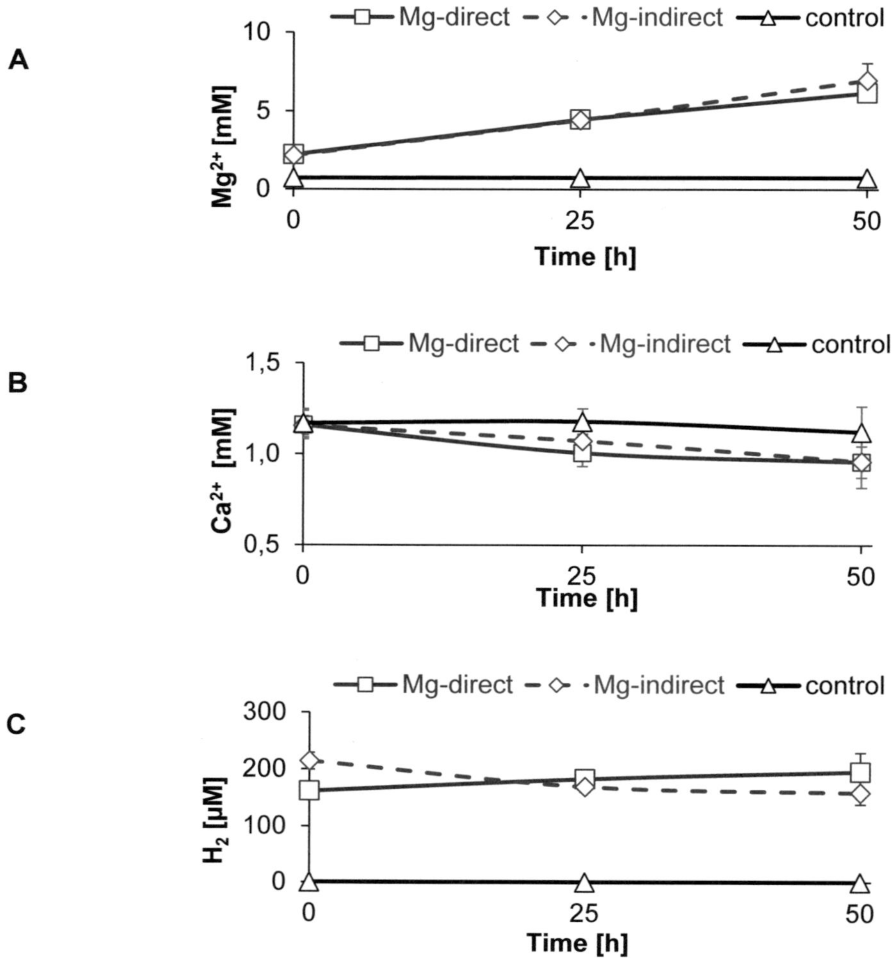

Figure 12. Mg^{2+} (A), Ca^{2+} (B) and H_2 (C) concentrations of the supernatants over time during HGF migration on pre-corroded Mg membrane (Mg-direct) and only affect by corroding Mg membrane in cell culture medium (Mg-indirect), control: cell culture medium. Data represent the mean±SD of three supernatants for each time point (n=3) [32].

3.6 Ca^{2+} storage in the corrosion layer of Mg

The Ca^{2+} analysis of the Mg membranes, obtained from both migration assays (Mg-direct, Mg-indirect) over time, showed an increased amount of Ca^{2+} in the Mg membrane as Ca^{2+} is incorporated in the corrosion layer of Mg (Fig.13). The Ca^{2+} content for the modified migration assay (Mg-direct) showed an increase over time from 1336±63 mg Kg^{-1} to 1931±82 mg Kg^{-1}. During the unmodified migration assay (Mg-indirect) the Ca^{2+} content was almost constant, ranging between 1256±56 and 1361±64 mg Kg^{-1}.

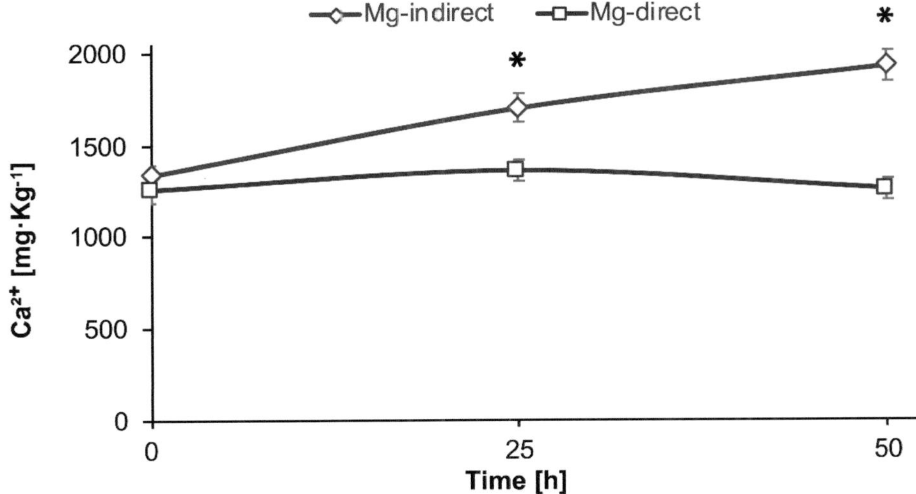

Figure 13. Ca^{2+} concentration of the corrosion layer on pre-corroded Mg membranes during cell migration assay over time with cells directly on Mg (Mg-direct) in comparison to Mg corrosion in a distance to cells (Mg-indirect). Each membrane was measured in triplicate, averaged and indicated as the mean±SD (n=3). Asterisks indicate significant differences between both assays (P<0.05) [32].

3.7 Effect of Mg^{2+} on cells

The effect of Mg^{2+} on cell migration, proliferation and viability of HGF was investigated using a scratch-based migration assay, a BrdU and a MTT assay, while using cell culture medium (CCM) supplemented with $MgCl_2 \cdot 6\ H_2O$. Medium with 25 mM $MgCl_2$ affected neither cell migration nor viability, while at 75 mM $MgCl_2$ the cell migration rate ($t_{50\%}$ = 38 h) was decreased compared to that of the control (CCM, $t_{50\%}$ = 13 h), but the viability was unaltered (Fig.14 A, C). The proliferation rate at 25 mM $MgCl_2$ was slightly increased (124±6) and at 75 mM was greatly decreased (12±7%), compared to that of the control (Fig.14 B).

To consider the increase of osmolality when $MgCl_2$ was added, controls with same osmolality were used, attained by adding mannitol to the CCM. The lower osmolality control (370 mOsmol Kg^{-1}) showed no effect on migration, proliferation and viability (Fig.14). The higher osmolality control (480 mOsmol Kg^{-1}) led to a decreased migration rate ($t_{50\%}$ = 20 h) without affecting cell proliferation and viability.

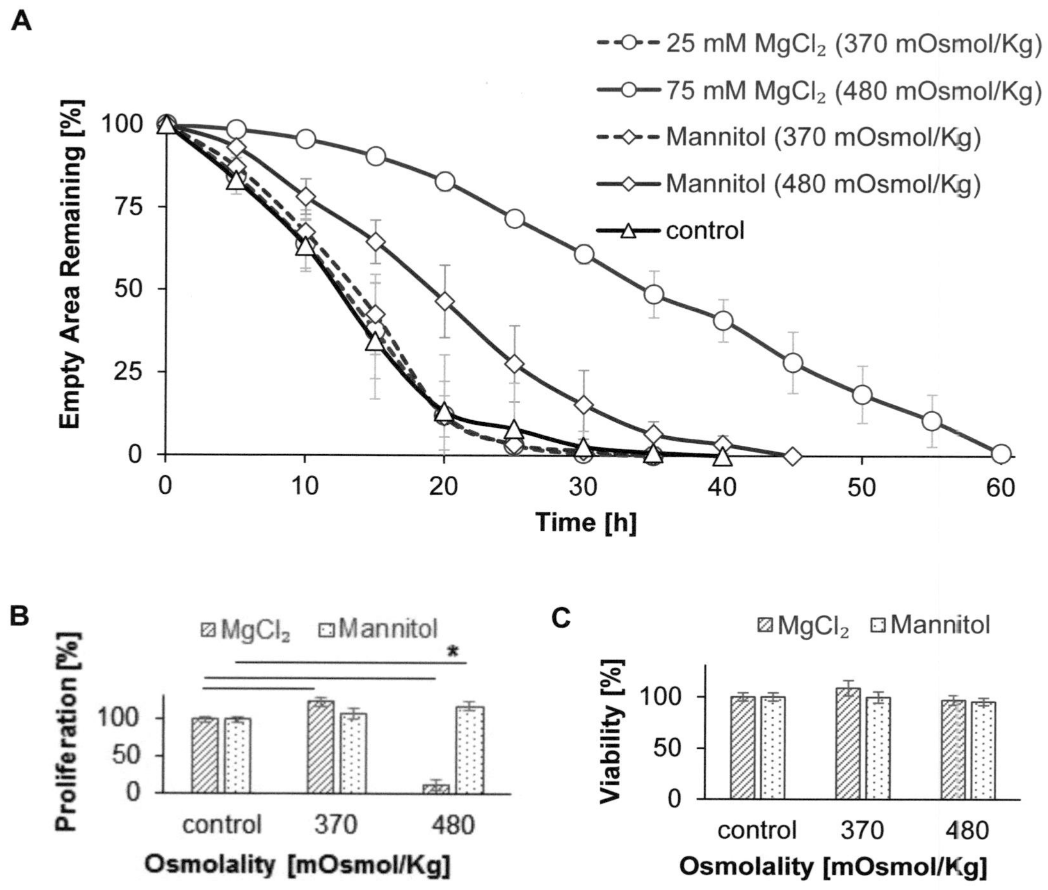

Figure 14. Effect of cell culture medium, supplemented with magnesium chloride (MgCl₂), on cell migration (A), proliferation (BrdU assay, B) and viability (MTT assay, C) of HGF, compared to same osmolalities adjusted with mannitol and cell culture medium (control). Test solutions for BrdU and MTT assay were incubated for 24 h, whereas during migration the cells were affected by the test solution until wound closure has been completed. Data represent the mean±SD (migration: n=4; proliferation/viability: n=5). Asterisks indicate significant differences (P<0.05) [32].

The increase in Cl⁻ by adding MgCl₂ was considered by using NaCl controls. Cell culture medium (CCM) with 50 mM NaCl has the same Cl⁻ concentration as CCM with 25 mM MgCl₂ and showed only a slight decrease in migration rate compared to 25 mM MgCl₂ and the control but did not affect cell proliferation and viability (Fig.15). At 150 mM NaCl, the migration rate was sharply decreased ($t_{50\%}$ = 29 h) compared to the control, but slightly increased compared to CCM with 75 mM MgCl₂ ($t_{50\%}$ = 38 h). While at 150 mM NaCl, the viability was not affected, the proliferation rate was sharply reduced in a similar way as the corresponding 75 mM MgCl₂ test solution (Fig.15 B, C).

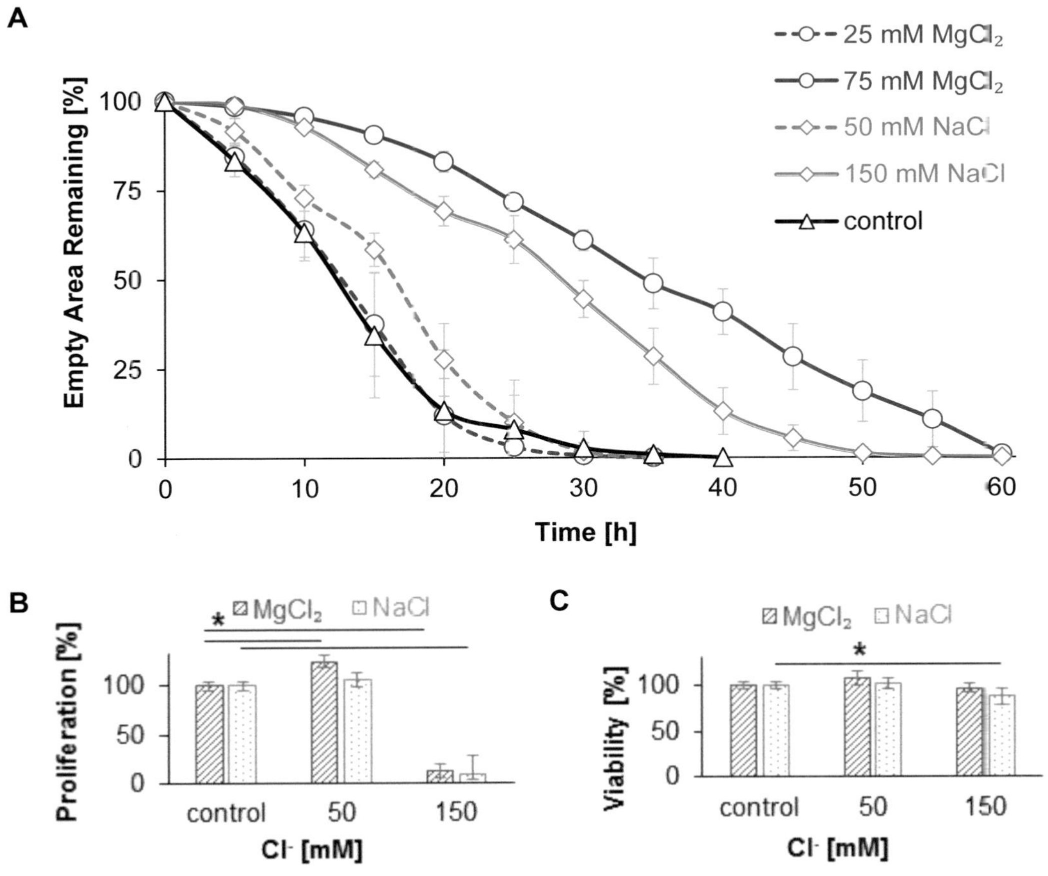

Figure 15. Effect of cell culture medium, supplemented with magnesium chloride (MgCl$_2$), on cell migration (A), proliferation (BrdU assay, B) and viability (MTT assay, C) of HGF, compared to same chloride concentration adjusted with sodium chloride and cell culture medium (control). Test solutions for BrdU and MTT assay were incubated for 24 h, whereas during migration the cells were affected by the test solution until wound closure has been completed. Data represent the mean±SD (migration: n=4; proliferation/viability: n=5). Asterisks indicate significant differences (P<0.05) [32].

To exclude the combined effect of osmolality increase and ionic strength increase when MgCl$_2$ was added, further test solutions containing additional mannitol and NaCl were prepared and tested in the assays. The 0 mM, 15 mM and 25 mM MgCl$_2$ test solutions all showed the same osmolality (450±3 mOsmol Kg^{-1}) and ionic strength (75 mM). All test solutions led to slightly slower migration rates compared to the control, but did not differ between each other (Fig.16 A). The cell proliferation and viability were not affected, since at all tested MgCl$_2$ concentrations the percentage values were over 80% compared to the control (Fig.16 B, C).

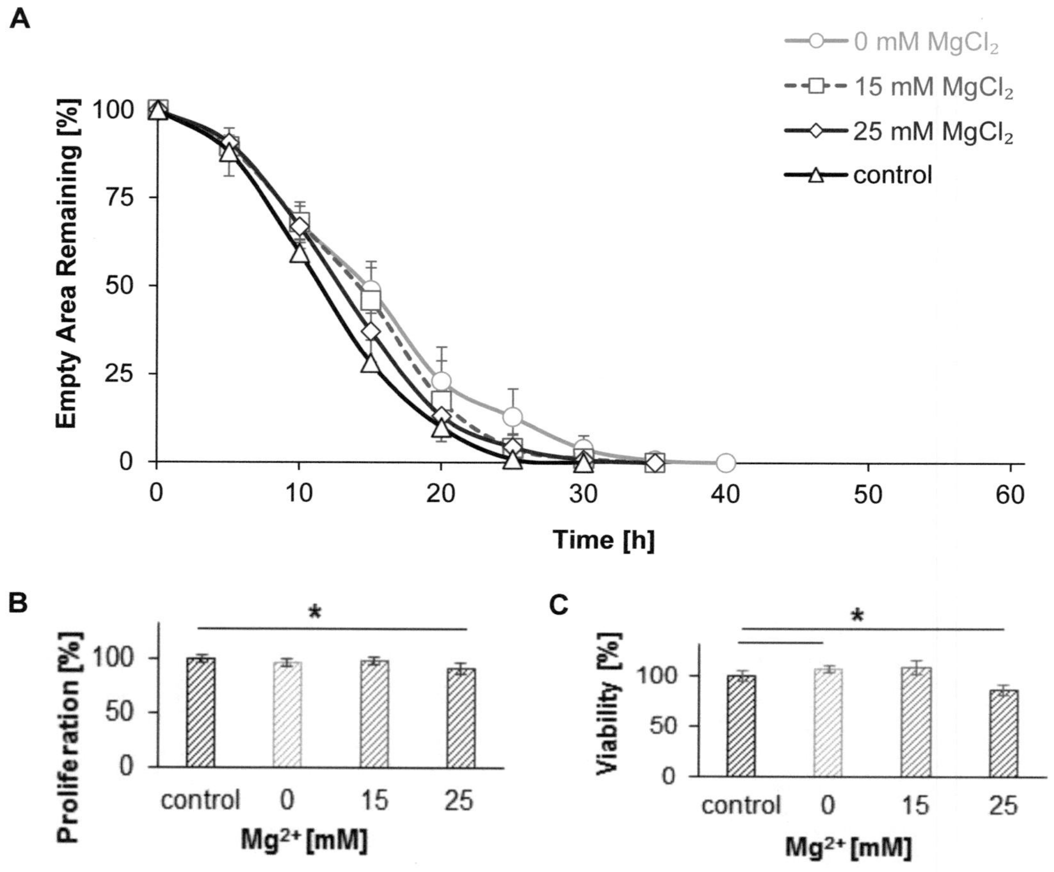

Figure 16. Effect of cell culture medium, supplemented with magnesium chloride (MgCl$_2$), sodium chloride (NaCl) and mannitol to reach the same osmolality (450 mOsmol Kg^{-1}) and ionic strength (75 Mm) on cell migration (A), proliferation (BrdU assay, B) and viability (MTT assay, C) of HGF, compared to cell culture medium (control). Test solutions for BrdU and MTT assay were incubated for 24 h, whereas during migration the cells were affected by the test solution until wound closure has been completed. Data represent the mean±SD (migration: n=4; proliferation/viability: n=6). Asterisks indicate significant differences (P<0.5) [32].

3.8 Effect of Ca^{2+} on cells

Calcium-free cell culture medium was supplemented with CaCl$_2$ · 2 H$_2$O and a migration, BrdU and MTT assays were performed to investigate the cell migration, proliferation and viability of HGF. The migration rate and the proliferation rate decreased with decreasing Ca^{2+} concentration (0.8 mM until 0 mM) (Fig.17 A, B). At 0 mM Ca^{2+} the time to close 50% of the initial cell-free area was 22 h. The control showed the fastest migration rate ($t_{50\%}$ = 13 h) and highest proliferation rate, which represented the highest Ca^{2+} concentration (1.16 mM) compared to the test solutions. All test solutions did not affect viability of HGF, as all percentage values were around 100% (Fig.17 C).

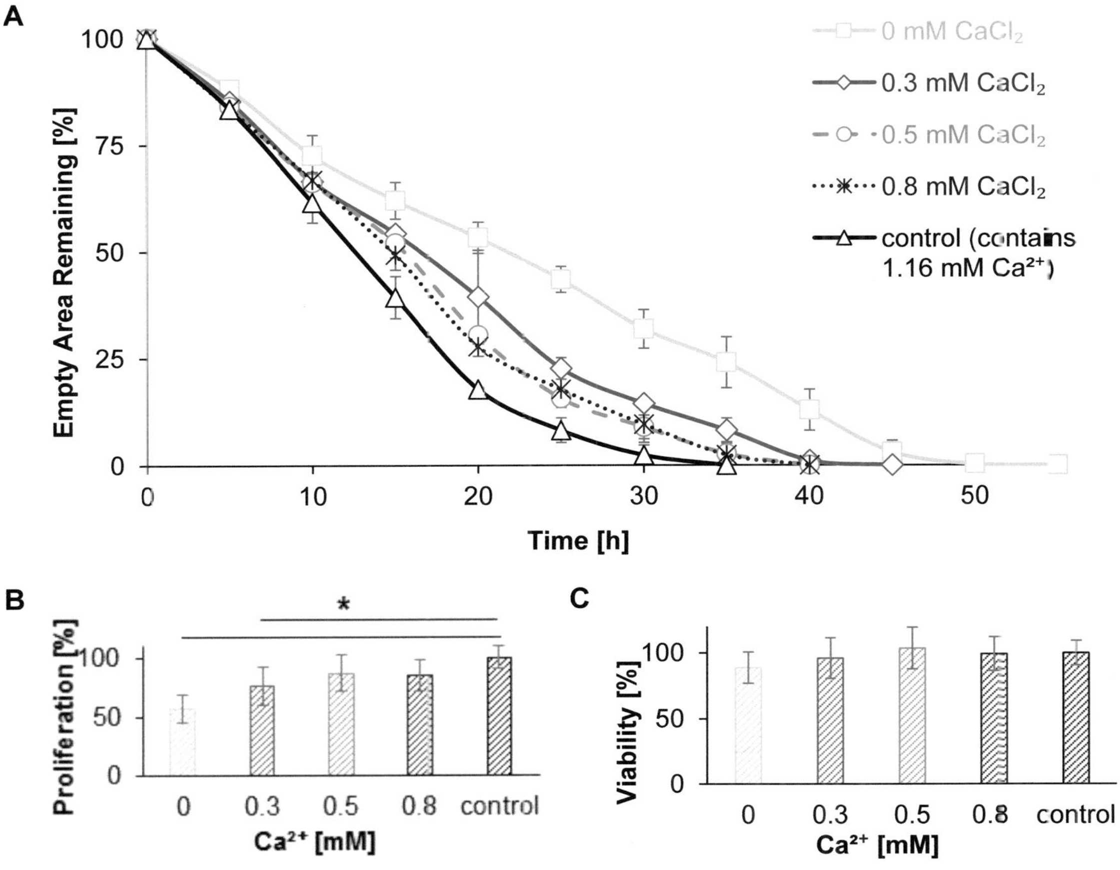

Figure 17. Effect of calcium free cell culture medium, supplemented with calcium chloride (CaCl₂) on cell migration (A), proliferation (BrdU assay, B) and viability (MTT assay, C) of HGF, compared to cell culture medium (control). Test solutions for BrdU and MTT assay were incubated for 24 h, whereas during migration the cells were affected by the test solution until wound closure has been completed. Data represent the mean±SD (migration: n=4; proliferation/viability: n=6). Asterisks indicate significant differences (P<0.5) [32].

3.9 Effect of H₂ on cells

Cell culture medium was enriched with H₂ (H-CCM) and two dilutions were prepared, which were all tested in the migration, BrdU and MTT assays to investigate cell migration, proliferation and viability of HGF. Neither the undiluted nor the diluted H-CCM affected migration rate, as well as proliferation and viability (Fig.18 A, C D). The initial H₂ concentration of undiluted (337±53 µM) and diluted H-CCM (123±6 µM and 46±3 µM) decreased over time and was no longer detectable after 60 min (Fig.18 B). Replacing the H-CCM every 30 min for the first 2.5 h affected neither proliferation nor viability of HGF (Fig.19).

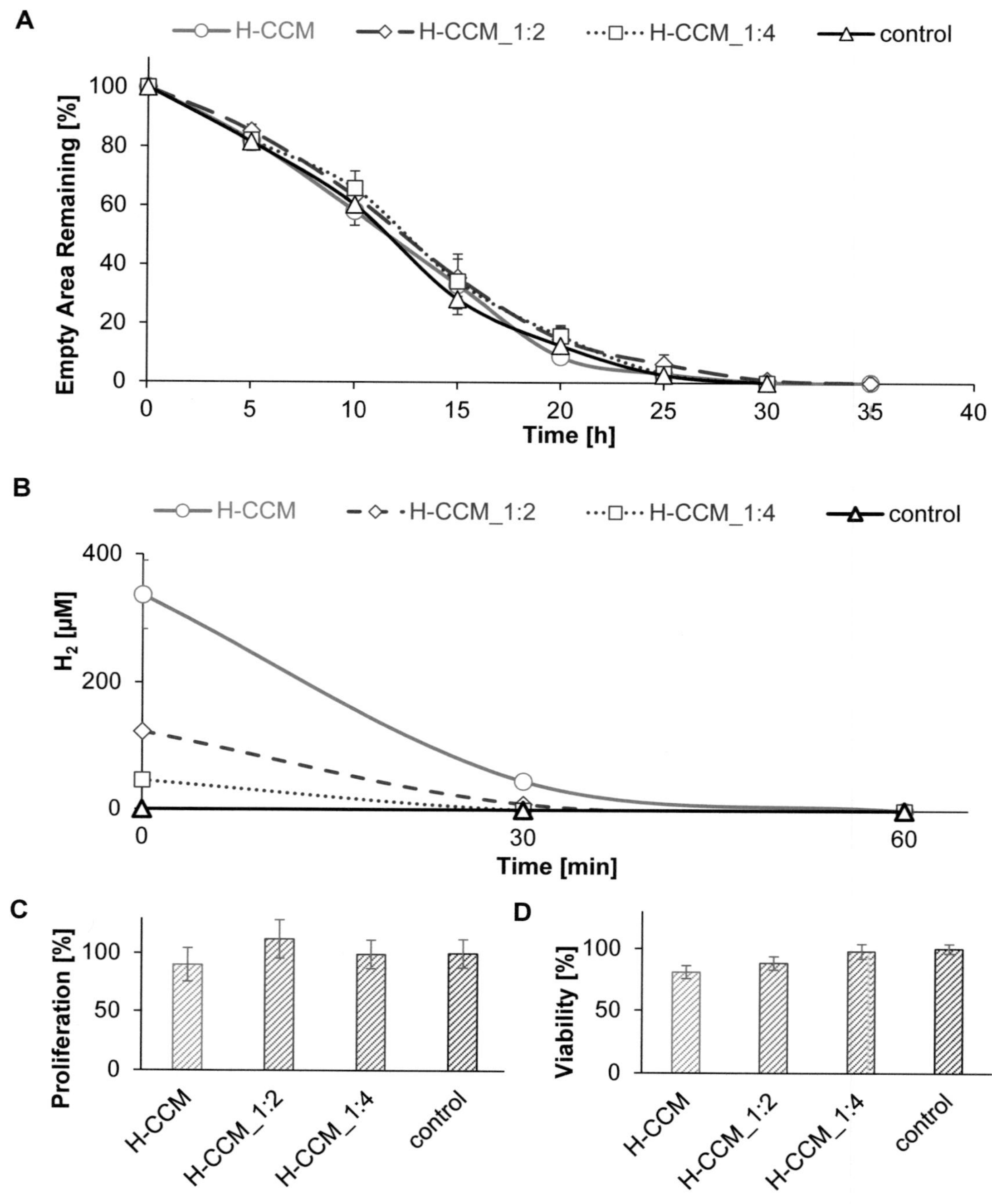

Figure 18. Effect of hydrogen-rich cell culture medium (H-CCM), generated with the Highdrogen Age2 Go for 23 min and two dilutions with CCM (1:2 and 1:4) on cell migration (A), proliferation (BrdU assay, C) and viability (MTT assay, D) of HGF compared to control (CCM). H-CCM for BrdU and MTT assay were incubated for 24 h, whereas during migration the cells were affected by H-CCM until wound closure has been completed. Data represent the mean±SD (migration: n=4; proliferation/viability: n=8). The H_2 concentration was measured for 30s at 0 h, after 0.5 h and after 1 h (C) during incubation under cell culture conditions and averaged (n=3). There is no significant difference between the H-CCM solutions and the control from the proliferation and viability assays. [32].

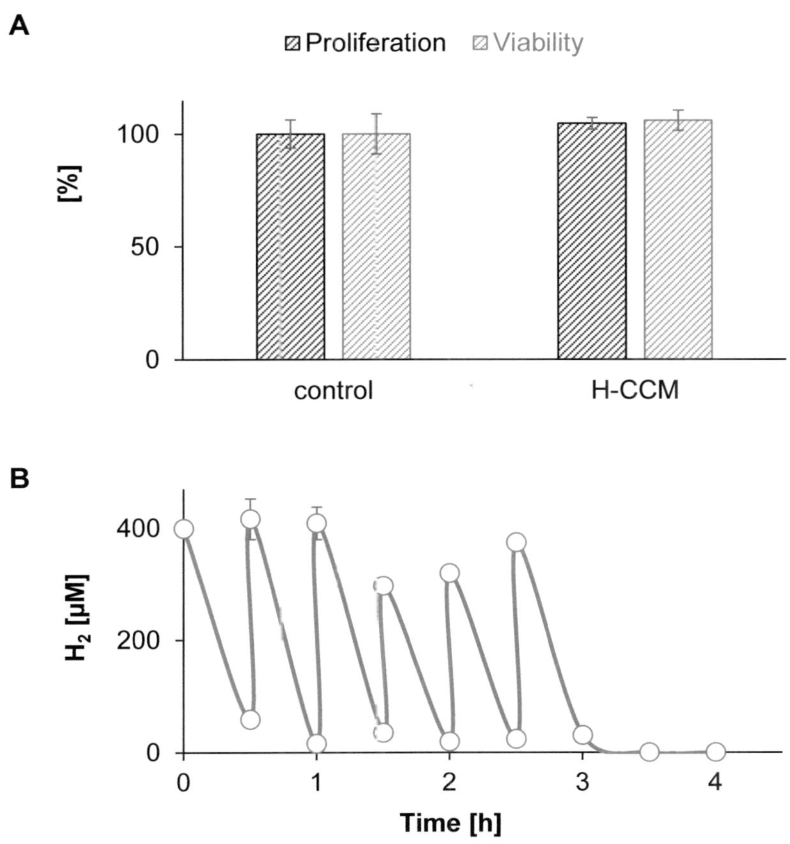

Figure 19. Effect of hydrogen-rich cell culture medium (H-CCM), generated with the Highdrogen Age$_2$ Go for 23 min, on cell proliferation (BrdU assay) and viability (MTT assay) of HGF compared to control (CCM) (A). H-CCM was renewed every 30 min over the first 2.5 h of the whole 24 h incubation time. Data represent the mean±SD (n=6). The H$_2$ concentration was measured for 30s (C) during incubation under cell culture conditions and averaged (n=3). There is no significant difference between the H-CCM and the control from the proliferation and viability assays [32].

3.10 Effect of Mg extracts on cells

The combined effect of all altered medium-related physical cues of corroding magnesium on HGF were analyzed by testing Mg extracts regarding cell migration, proliferation and viability. The Mg extracts were obtained by immersing the magnesium membrane in cell culture medium (CCM) with (w CO$_2$) and without (w/o CO$_2$) carbon dioxide exchange. Incubation with Mg-extract (w/o CO$_2$) led to a slightly slower migration rate ($t_{50\%}$ = 15 h) compared to the control (Fig.20 A), without affecting cell proliferation and viability (Fig.20 B, C). In contrast, when cells were incubated with Mg-extract (w CO$_2$), no migration rate was detectable as the cells became rounded after 5 h and remained in their position over the whole assay time. Additionally, the proliferation and viability were below 75% compared to the control.

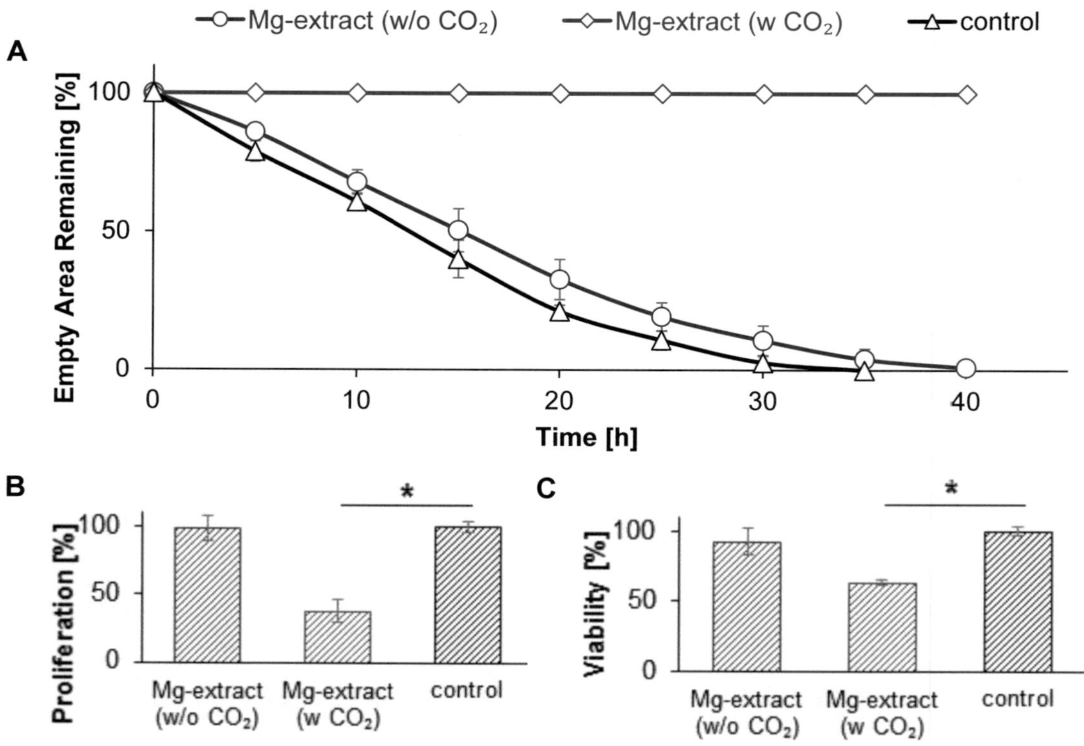

Figure 20. Effect of Mg extracts, received by corrosion of the Mg membrane (10x10mm), 140 µm thickness) in 10 mL cell culture medium with and without CO_2 exchange (w CO_2) and without (w/o CO_2) on cell migration (A), proliferation (B) and viability (C) of HGF. Mg extracts for BrdU and MTT assay were incubated for 24 h, whereas during migration the cells were affected by Mg extracts until wound closure has been completed. Data represent the mean±SD (migration: n=4; proliferation/viability: n=4). Asterisks indicates significant differences (P<0.05) [32].

The measured H_2, Mg^{2+} and Ca^{2+} concentrations of both Mg extracts over time are presented in Figure 21. The H_2 concentration of both extracts was, negligibly small, with values below 4 µm directly after preparing the extracts and was not detectable anymore after 15 min. The Mg-extract (w CO_2) revealed an extremely elevated Mg^{2+} concentration (58±1 mM) and a reduced Ca^{2+} concentration (0.57±0.01 mM) compared to the Mg-extract (w/o CO_2) and to the control. The Mg^{2+} concentration of the Mg-extract (w/o CO_2) was 4.4±0.01 mM and the Ca^{2+} concentration was 0.93±0.02 mM.

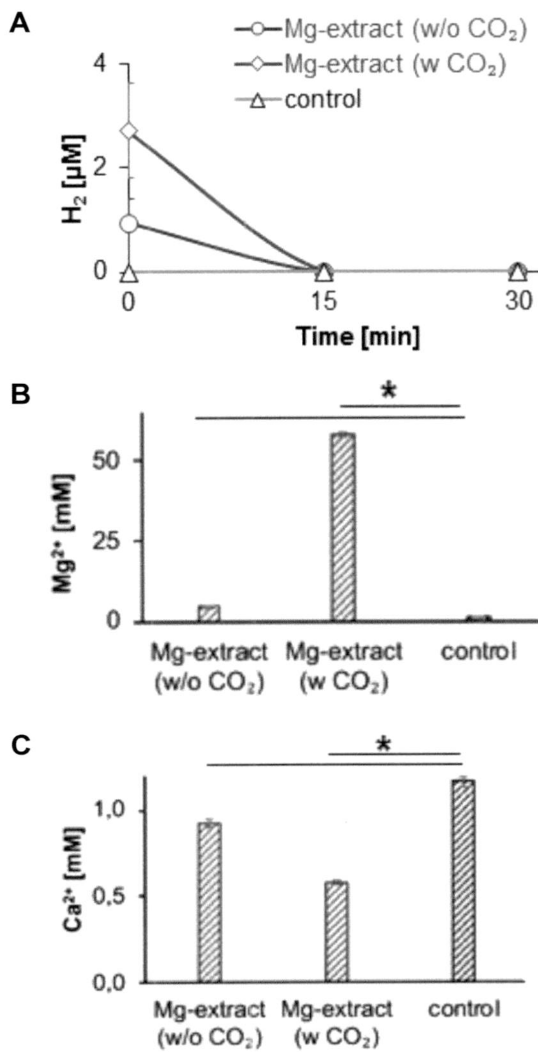

Figure 21. The H_2 (A), Mg^{2+} (B) and Ca^{2+} (C) concentration of the Mg-extracts (w/o CO_2, w CO_2), which were tested in the migration, proliferation and viability assay. The Mg^{2+} and Ca^{2+} concentrations were measured in triplicate of each extract, averaged and indicated as the mean±SD of three independent prepared extracts (n=3). The H_2 concentration was measured for 30 s, averaged and indicated as the mean±SD of three independent prepared extracts (n=3). Asterisks indicate significant differences (P<0.05) [32].

4 Discussion

Magnesium-based implants are very promising in application for dentistry, due to their excellent biocompatibility and biodegradability. Thus, barrier membranes made of pure magnesium can be used for Guided Bone Regeneration (GBR) procedures [1, 4]. Due to the often reported membrane exposure associated with utilized non-resorbable membrane materials made of titanium, cell migration of human gingival fibroblasts (HGF) on the barrier membrane is crucial to close the exposed membrane and to ensure optimal healing success [8]. Cell migration of HGF on magnesium membranes was already investigated *in vitro* and *in vivo* during my previous research [4]. However, there is no comparison to membrane materials that are currently used. Regarding cell migration studies on magnesium surfaces, we must consider that magnesium degrades by forming magnesium ions, molecular hydrogen and hydroxide ions [1]. Furthermore, the surface is altered due to formation of a corrosion layer [12]. All parameters affecting cell behaviour, including cell migration, viability and proliferation are called 'physical cues'. The effect of physical cues on cell behaviour of HGF has not been investigated in detail. Identification of possible effects, especially on HGF migration, including specific concentration of the medium-altered physical cues, will provide knowledge to design the targeted magnesium corrosion rate for optimal healing success.

In this study, the migration behaviour of HGF on titanium surfaces was analyzed, including surface measurements to characterize the surface wettability, roughness and topography, and to compare these results with existing results for magnesium. To prove our assumption that medium alterations caused by magnesium corrosion have a major impact on cell migration rather than surface alterations, the migration assay was modified to exclude the surface impact. As a main part of this study, the effect of the different medium-altered physical cues on cell migration, proliferation and viability of the HGF was analyzed to explain the migration behaviour on magnesium membranes. Our results showed that cell migration on magnesium surfaces is 1.7 fold slower compared to tissue culture plastic and 2.8 fold slower compared to titanium. We demonstrated that migration rate is mainly affected by medium alterations, while the surface has only a minor effect, verified by performing the modified migration assay. More specifically, we identified the Mg^{2+}/Ca^{2+} ratio and the constantly elevated H_2 level as parameters that seem to explain the slow migration behaviour of HGF on magnesium membranes.

Surface measurements yielded roughness values below 200 nm for magnesium (pre-corroded/ uncorroded), titanium and tissue culture plastic. We did not find a correlation between migration rate and roughness. According a study of Kim *et al.* [33] roughness has no influence on cell attachment of HGF in the range of 200 nm or less. Accordingly, HGF attachment is mainly affected by hydrophilicity with highest fibroblast adhesion and proliferation at a water contact angle (WCA) between 50° and 60° [34]. As already known, high cell attachment leads to a slower migration rate [35]. The comparison to the control surfaces shows that cell migration is faster on titanium than on tissue culture plastic surfaces. This can be explained by the lower hydrophilicity of the tissue culture plastic surface, which is in the stated range for highest fibroblast attachment and slower cell migration. Magnesium showed almost the same WCA as titanium, but the HGF migration was significantly reduced. Therefore, we assumed that the surface condition of magnesium does not determine cell migration on magnesium membranes.

Immunohistological investigations of cellular structure showed that HGF treated with medium containing 75 mM $MgCl_2$ lead to increased focal adhesions, showing that the Mg^{2+} concentration is one physical cue that affects cell adhesion as a crucial parameter for cell migration. The modified migration assay allowed to exclude all surface impacts to ensure that the cells were only affected by the dissolved corrosion products of magnesium. Our results showed that cell migration of each assay (modified/ unmodified) is comparable. Therefore we demonstrated that medium alterations caused by magnesium corrosion are the main reason for slower migration rate on magnesium membranes. Moreover, the characterized supernatants over time during cell migration showed similar Mg^{2+}, Ca^{2-} and H_2 concentrations for both assays.

To obtain more precise results of how each medium-altered physical cue affects cell migration, cell culture medium supplemented with Mg and Ca salts, as well as H_2 enriched medium was tested. Moreover, the combined effect of single physical cues was investigated by preparing different magnesium extracts.

Cell culture medium supplemented with magnesium chloride showed no effects on HGF migration, proliferation or viability up to 25 mM. The cell migration and prolifera-tion were sharply reduced only at 75 mM. We could exclude effects of increasing osmolality and chloride concentrations when magnesium chloride was added by using adequate controls. By comparing this result with the determined Mg^{2+} concentration during the cell migration assay on magnesium membranes we could exclude the Mg^{2+} concentration as a single potential physical cue affecting migration rate, since the Mg^{2+}

concentration during the migration assay was below 25 mM. We only measured the concentration of the whole supernatant; therefore, we cannot exclude the possibility that the Mg^{2+} concentration directly on the Mg surface may have been slightly higher. Increasing the calcium chloride concentration from 0 mM to 0.8 mM in the medium by using calcium-free cell culture medium lead to an increased migration rate. These results are in line with the literature as Ca^{2+} is required for cell detachment resulting in faster cell migration [36, 37]. During HGF migration on Mg surfaces, the Ca^{2+} concentration of the supernatant decreased only slightly from 1.16 mM to 0.95 mM. The Ca^{2+} decrease can be explained by $Ca_5(PO_4)(OH)$ and $Ca_3(PO_4)_2$ precipitation in the corrosion layer [38], which could be confirmed by ICP-OES in our studies. Combining both results, we assume, that the slightly Ca^{2+} decrease during cell migration on Mg surfaces only makes a very small contribution to the slow migration rate of HGF.

Furthermore, the effect of H_2 enriched cell culture medium on cell behaviour of HGF was tested. Our results showed that initial concentrations up to 336 µM had no effect on cell migration, proliferation and viability. Due to the rapid clearing, the H_2 of the medium was completely eliminated after 60 min. Therefore, we could not ensure conditions identical to our established migration assay on magnesium surfaces having constant H_2 concentration of 180 µM over 50 h. According to the literature HGF even react after short exposure of 2 h to H_2 enriched cell culture medium (690 µM) as cell migration and proliferation was promoted resulting in faster wound healing *in vitro* [39]. Therefore, we assumed that HGF only react at higher H_2 concentration.

Finally, we tested two different prepared magnesium extracts (with and without CO_2) to investigate the combined effect of all medium-altered physical cues on HGF migration. Both prepared Mg extracts showed negligibly small initial H_2 concentrations of 4 µM.

Magnesium extract without CO_2 exchange lead to a slightly reduced cell migration rate, although the Mg^{2+} concentration was only slightly increased (4.5 mM) and the Ca^{2+} concentration slightly decreased (0.93 mM). As we cannot explain the reduced migration rate with the slightly increased Mg^{2+} concentration, we can conclude that the decrease of Ca^{2+} might affect migration rate. The effect becomes more obvious at tested Mg extract with CO_2, as there was no cell migration detectable, although the Mg^{2+} concentration was below 75 mM. In this concentration range, cell migration might be possible even in a reduced way.

These results revealed that it s not the absolute concentration of both ions that determines migration behaviour, but that the ratio between Mg^{2+} and Ca^{2+} is more crucial, which can be supported by literature [36, 37, 40]. The Mg^{2+}/Ca^{2+} ratio of the Mg extract with CO_2 exchange was 102 and without CO_2 exchange was 5 (for comparison: control = 0.6).

By applying this knowledge to the established migration assay on magnesium, we showed an increase in the Mg^{2+}/Ca^{2+} ratio from 1.8 at the beginning to 4.4 after 25 h and to 7.4 after 50 h. Interestingly, the ratio after 25 h was almost the same as the ratio of the Mg extract without CO_2. However, the migration rate on Mg membranes s slower compared to the Mg extract without CO_2. The only difference is that, during migration on Mg membranes, the H_2 concentration is constantly 180 µM; during migration incubating with Mg extract without CO_2, the H_2 concentration remained at zero after 15 min for the whole migration assay period. Therefore, we assume, that not only the Mg^{2+}/Ca^{2+} ratio affects migration behaviour but that the combination of Mg^{2+}/Ca^{2+} ratio and dissolved H_2 in the cell culture medium lead to reduced migration rate on Mg surfaces. This study provides primary results of migration studies of HGF on Mg membranes, which should resemble the migration behaviour of HGF during the initial hours of Mg implantation *in vivo*. However, this study has potential limitations. In order to get more reliable results the analysis should be performed with more replicates. Furthermore, the applicability of *in vitro* experiments to *in vivo* situations is very debatably, as the *in vitro* conditions can not completely simulate the *in vivo* situations. However, with the knowledge of this study the environmental conditions during the application of Mg membranes in GBR procedure can be changed in order to ensure faster wound healing.

5 Literature

[1] Zhao D, Witte F, Lu F, Wang J, Li J, Qin L. Current status on clinical applications of magnesium-based orthopaedic implants: A review from clinical translational perspective. Biomaterials 2017;112:287-302.

[2] Sissi C, Palumbo M. Effects of magnesium and related divalent metal ions in topoisomerase structure and function. Nucleic Acids Research 2009;37:702-11.

[3] Waizy H, Seitz J-M, Reifenrath J, Weizbauer A, Bach F-W, Meyer-Lindenberg A, Denkena B, Windhagen H. Biodegradable magnesium implants for orthopedic applications. Journal of Materials Science 2013;48:39-50.

[4] Amberg R, Elad A, Rothamel D, Fienitz T, Szakacs G, Heilmann S, Witte F. Design of a migration assay for human gingival fibroblasts on biodegradable magnesium surfaces. Acta Biomaterialia 2018.

[5] Liu J, Kerns DG. Mechanisms of Guided Bone Regeneration: A Review. The Open Dentistry Journal 2014;8:56-65.

[6] Robinson DA, Griffith RW, Shechtman D, Evans RB, Conzemius MG. In vitro antibacterial properties of magnesium metal against Escherichia coli, Pseudomonas aeruginosa and Staphylococcus aureus. Acta Biomater 2010;6:1869-77.

[7] Burmester A, Willumeit-Romer R, Feyerabend F. Behavior of bone cells in contact with magnesium implant material. J Biomed Mater Res B Appl Biomater 2017;105:165-79.

[8] Zitzmann N, Naef R, Schärer P. Resorbable Versus Nonresorbable Membranes in Combination with Bio-Oss for Guided Bone Regeneration1997.

[9] Lang NP, Hämmerle CHF, Brägger U, Lehmann B, Nyman SR. Guided tissue regeneration in jawbone defects prior to implant placement. Clinical Oral Implants Research 1994;5:92-7.

[10] Behring J, Junker R, Walboomers XF, Chessnut B, Jansen JA. Toward guided tissue and bone regeneration: morphology, attachment, proliferation, and migration of cells cultured on collagen barrier membranes. A systematic review. Odontology 2008;96:1-11.

[11] Li L, Zhang M, Li Y, Zhao J, Qin L, Lai Y. Corrosion and biocompatibility improvement of magnesium-based alloys as bone implant materials: a review. Regenerative Biomaterials 2017.

[12] Charyeva O, Feyerabend F, Willumeit R, Zukowski D, Gasqueres C, Szakács G, Ahmad Agha N, Hort N, Gensch F, Cecchinato F, Jimbo R, Wennerberg A, Lips K. In Vitro Resorption of Magnesium Materials and its Effect on Surface and Surrounding Environment2015.

[13] Günzler H, Bahadir AM, Danzer K, Engewald W, Fresenius W, Galensa R, Huber W, Linscheid M, Schwedt G, Tölg G. Analytiker-Taschenbuch: Springer Berlin Heidelberg; 2013.

[14] Ul-Hamid A. A Beginners' Guide to Scanning Electron Microscopy: Springer International Publishing; 2018.

[15] Michler GH. Electron Microscopy of Polymers: Springer Berlin Heidelberg; 2003.

[16] Emter E. Literatur und Quartentheorie: die Rezeption der modernen Physik in Schriften zur Literatur und Philosophie deutschprachiger Autoren (1925-1970): De Gruyter; 1995.

[17] Schatten H. Scanning Electron Microscopy for the Life Sciences: Cambridge University Press; 2013.

[18] Flegler SL, Heckman JW, Klomparens KL. Elektronenmikroskopie: Grundladen-Methoden-Anwendungen. Heidelberg: Spektrum Akademischer Verlag; 1995.

[19] Braga PC, Ricci D. Atomic Force Microscopy: Biomedical Methods and Applications: Humana Press; 2004.

[20] Halevi G. Process and Operation Planning: Revised Edition of The Principles of Process Planning: A Logical Approach: Springer Netherlands; 2003.

[21] Haas W. Oberflächenbeurteilung- Rauheitsmessung. In: Universität, editor. Stuttgart2012. p. 1-18.

[22] Stout KJ. Development of Methods for the Characterisation of Roughness in Three Dimensions: Penton Press; 2000.

[23] Bellitto V. Atomic Force Microscopy: Imaging, Measuring and Manipulating Surfaces at the Atomic Scale: IntechOpen; 2012.

[24] Voigtländer B. Scanning Probe Microscopy: Atomic Force Microscopy and Scanning Tunneling Microscopy: Springer Berlin Heidelberg; 2015.

[25] Morita S, Giessibl FJ, Wiesendanger R. Noncontact Atomic Force Microscopy: Springer Berlin Heidelberg; 2009.

[26] Eaton P, West P. Atomic Force Microscopy: OUP Oxford; 2010.

[27] Berg J. Wettability: CRC Press; 1993.

[28] Owens DK, Wendt RC. Estimation of the surface free energy of polymers. Journal of Applied Polymer Science 1969;13:1741-7.

[29] Kaelble D. Dispersion-polar surface tension properties of organic solids. The Journal of Adhesion 1970;2:66-81.

[30] Pan C-J, Pang L-Q, Hou Y, Lin Y-B, Gong T, Liu T, Ye W, Ding H-Y. Improving Corrosion Resistance and Biocompatibility of Magnesium Alloy by Sodium Hydroxide and Hydrofluoric Acid Treatments. Applied Sciences 2016;7:33.

[31] Mattox DM. Chapter 2 - Substrate ("Real") Surfaces and Surface Modification. In: Mattox DM, editor. Handbook of Physical Vapor Deposition (PVD) Processing (Second Edition). Boston: William Andrew Publishing; 2010. p. 25-72.

[32] Amberg R, Elad A, Beuer F, Vogt C, Bode J, Witte F. Effect of physical cues of altered extract media from biodegradable magnesium implants on human gingival fibroblasts. Acta Biomaterialia 2019.

[33] Kim YS, Shin SY, Moon SK, Yang SM. Surface properties correlated with the human gingival fibroblasts attachment on various materials for implant abutments: a multiple regression analysis. Acta odontologica Scandinavica 2015;73:38-47.

[34] Kim SH, Ha HJ, Ko YK, Yoon SJ, Rhee JM, Kim MS, Lee HB, Khang G. Correlation of proliferation, morphology and biological responses of fibroblasts on LDPE with different surface wettability. J Biomater Sci Polym Ed 2007;18:609-22.

[35] DiMilla PA, Barbee K, Lauffenburger DA. Mathematical model for the effects of adhesion and mechanics on cell migration speed. Biophysical Journal 1991;60:15-37.

[36] Lange TS, Bielinsky AK, Kirchberg K, Bank I, Herrmann K, Krieg T, Scharffetter-Kochanek K. Mg2+ and Ca2+ Differentially Regulate β1 Integrin-Mediated Adhesion of Dermal Fibroblasts and Keratinocytes to Various Extracellular Matrix Proteins. Experimental Cell Research 1994;214:381-8.

[37] Lange TS, Kirchberg K, Bielinsky AK, Leuker A, Bank I, Ruzicka T, Scharffetter-Kochanek K. Divalent cations (Mg2+, Ca2+) differentially influence the β1 integrin-mediated migration of human fibroblasts and keratinocytes to different extracellular matrix proteins. Experimental dermatology 1995;4:130-7.

[38] Willumeit R, Fischer J, Feyerabend F, Hort N, Bismayer U, Heidrich S, Mihailova B. Chemical surface alteration of biodegradable magnesium exposed to corrosion media. Acta Biomater 2011;7:2704-15.

[39] Xiao L, Miwa N. Hydrogen-rich water achieves cytoprotection from oxidative stress injury in human gingival fibroblasts in culture or 3D-tissue equivalents, and wound-healing promotion, together with ROS-scavenging and relief from glutathione diminishment. Human Cell 2017;30:72-87.

[40] Grzesiak JJ, Pierschbacher MD. Shifts in the concentrations of magnesium and calcium in early porcine and rat wound fluids activate the cell migratory response. The Journal of clinical investigation 1995;95:227-33.

List of Publications

Amberg R., Elad A., Rothamel D., Fienitz T., Szakacs G., Heilmann S., Witte F., Design of a migration assay for human gingival fibroblasts on biodegradable magnesium surfaces, Acta Biomaterialia, October 2018.

Amberg R., Elad A., Beuer F., Vogt C., Bode J., Witte F., Effect of physical cues of altered extract media from biodegradable magnesium implants on human gingival fibroblasts, Acta Biomaterialia, July 2019.

Acknowledgement

I would like to express my sincere gratitude to my first supervisor, Prof. Dr. Frank Witte, for the continuous support, motivation, patience and immense knowledge during the completion of my PhD and related research. Furthermore, I would like to thank my second supervisor, Prof. Dr. Florian Beuer, for his encouragement and practical insights from the clinical perspective, which had given me a deeper understanding of the clinical application of my research. I would also like to thank all members of my working group for the stimulating discussions, their meaningful support and the pleasant working atmosphere. Last but not least, I thank my family: my mother, brother and both sisters for supporting me spiritually throughout writing my thesis.

Christoph Behnke

Gibt es Lockerbraunerden im Nordschwarzwald?

Behnke, Christoph: Gibt es Lockerbraunerden im Nordschwarzwald?. Hamburg, Bachelor + Master Publishing 2015

Originaltitel der Abschlussarbeit: Gibt es Lockerbraunerden im Nordschwarzwald?

Buch-ISBN: 978-3-95820-288-7
PDF-eBook-ISBN: 978-3-95820-788-2
Druck/Herstellung: Bachelor + Master Publishing, Hamburg, 2015
Covermotiv: © Kobes · Fotolia.com
Zugl. Johann Wolfgang Goethe-Universität Frankfurt am Main, Frankfurt am Main, Deutschland, Bachelorarbeit, Juli 2013

Bibliografische Information der Deutschen Nationalbibliothek:
Die Deutsche Nationalbibliothek verzeichnet diese Publikation in der Deutschen Nationalbibliografie; detaillierte bibliografische Daten sind im Internet über http://dnb.d-nb.de abrufbar.

© Bachelor + Master Publishing, Imprint der Diplomica Verlag GmbH
Hermannstal 119k, 22119 Hamburg
http://www.diplomica-verlag.de, Hamburg 2015
Printed in Germany

Danksagung

Mein Dank gilt vor allem Herrn Prof. Dr. Heinrich Thiemeyer sowohl für die Übernahme der Funktion des Erstgutachters, wie auch für die Begleitung bei einer ersten Übersichtsbegehung im Gelände und den fachlichen Diskussionen. Neben der Auswertung der Schwermineralanalyse weiß ich seine stete Unterstützung egal zu welcher Zeit über den gesamten Arbeitsprozess hinweg und den Mehrwert, welcher sich dadurch für meine bodenkundliche Ausbildung ergibt, sehr zu schätzen.

Daneben danke ich Herrn Dr. Rainer Dambeck, der mir ebenso in seiner Tätigkeit als Zweitgutachter dieser Arbeit immer fachlichen Beistand leistete.

Weiterhin gebührt mein Dank dem Laborpersonal Frau Doris Bergmann-Dörr und Frau Dagmar Schneider sowie den zahlreichen Hilfswissenschaftlern, welche durch ihre Anleitung und Begleitung der Laborarbeiten mit zum Gelingen dieser Arbeit beigetragen haben.

Inhaltsverzeichnis

Abbildungsverzeichnis

Abkürzungsverzeichnis

<	kleiner als		gU	Grobschluff
>	größer als		Hrsg.	Herausgeber
%	von Hundert		Jh.	Jahrhundert
°C	Grad Celsius		km	Kilometer
µm	Mikrometer		km²	Quadratkilometer
II	zweite periglaziale Lage		LB	Basislage
a	Jahre		LH	Hauptlage
AAS	Atom-Absorptionsspektrometer		Ls3	mittelsandiger Lehm
Abb.	Abbildung		Ls4	stark sandiger Lehm
Al	Aluminium		LST	Laacher-See-Tephra
Al_d	dithionitlösliches Aluminium		Mio.	Millionen
Al_{ges}	pedogenes Aluminium gesamt		ml	Milliliter
Al_o	oxalatlösliches Aluminium		mm	Millimeter
BK	Bodenkundliche Karte		Mn	Mangan
bzw.	beziehungsweise		Mn_d	dithionitlösliches Mangan
cm	Zentimeter		Mn_{ges}	pedogenes Mangan gesamt
cm³	Kubikzentimeter		Mn_o	oxalatlösliches Mangan
C_{org}	organischer Kohlenstoff		mS	Mittelsand
E	East (= Ostwert)		mU	Mittelschluff
erw.	erweiterte		N	North (= Nordwert)
et al.	et altera (= und weitere)		NHN	Normal-Höhennull
f.	folgende		Nr.	Nummer
Fe	Eisen		Sl3	mittellehmiger Sand
Fe_d	dithionitlösliches Eisen		Sl4	stark lehmiger Sand
Fe_{ges}	pedogenes Eisen gesamt		Slu	schluffig-lehmiger Sand
Fe_o	oxalatlösliches Eisen		SWR	Südwest-Rundfunk
ff.	fortfolgende		T	Ton
fS	Feinsand		TK	Topographische Karte
fU	Feinschluff		überarb.	überarbeitet
g	Gramm		vgl.	vergleiche
Gr	Grus		Wf	Feinwurzelanteil
gS	Grobsand		Wg	Grobwurzelanteil

1 Einleitung

Der Boden bildet die Schnittstelle aller natürlichen Kompartimente, innerhalb derer Wechselwirkungen von Elementen der Lithosphäre, der Hydrosphäre, der Atmosphäre wie auch der Biosphäre stattfinden.

Aufgrund seiner Funktion als Speichermedium und Versorger pflanzlichen Lebens stellt der Boden ein esentielles Element im Leben der Menschen dar, welches eine Sesshaftwerdung aufgrund gesicherter Nahrungsgrundlage und damit einhergehend im weiteren die Herausbildung von Zivilisation und Kultur ermöglichte

Diese Lebensgrundlage gilt es zu bewahren. Doch erst durch ein Verständnis für die Prozesse, welche sich an der Pedogenese beteiligt zeigen, ist es möglich auf diese lebenwichtige wie jedoch ebenso knapp bemessene Ressource einwirkende Mechanismen auf ihr Schadenspotential zu bemessen und in der Folge einen entsprechenden Schutz zu gewährleisten.

Diese Ausarbeitung soll entsprechend einen geringen Teil zum Grundverständnis pedogenetischer Prozesse in Böden darstellen.

1.1 Anlass und Ziel

Im Bereich des Nordschwarzwaldes konnte speziell an den Hängen von Erhebungen des Grindenschwarzwaldes bei früheren Geländebegehungen eine hohe Diversität an Subtypen der Braunerde festgestellt werden, unter welchen sich insbesondere ein Boden mit sehr locker gelagerten Oberbodenhorizonten und hoher Tixotrophie bei der Fingerprobe von den zu erwartenden wie auch generalisiert kartierten Bodentypen abgrenzen ließ (vgl. BÜK CC 7910 Freiburg-Nord).

Im Zuge dieser Ausarbeitung sollen ausgewählte Profile des fraglichen Bodentyps in der Region nördlich der Hornisgrinde im Nordschwarzwald eingangs den zumeist verbreiteter Braunerdesubtypen entgegengestellt und über feldbodenkundliche wie auch im Labor erhobene Daten verglichen werden, um signifikante Unterscheidungsmerkmale, welche einen möglichen Aufschluss über die Pedogenese dieser Böden liefern könnten, herauszuarbeiten. Anschließend findet ein Vergleich der fraglichen Bodenprofile über das Erscheinungsbild und die erhobenen Werte mit beschriebenen Bodenprofilen ähnlicher Beschaffenheit und vergleichbarem Ausbreitungsmuster statt, sodass sich auf Grundlage dieser eine Zuordnung innerhalb der Bodenklassifikation mit anschließender Diskussion über die Genese dieses Bodentyps ergibt.

1.2 Aktueller Kenntnisstand

Erstmalig wurde durch SCHÖNHALS im Jahre 1957 ein in den Hochlagen der deutschen Mittelgebirge sowie an den Unterhängen der Alpen vorzufindender Subtyp der Braunerden beschrieben, welcher sich insbesondere durch eine ockerbraune Färbung des Oberbodens, eine ungewöhnlich hohe

Lockerheit des Gefüges sowie trotz einer niedrigen Basensättigung keine Anzeichen einer Podsolierung aufweisend auszeichnet, dessen Genese er auf ein lössartiges Gestein der Jüngeren Tundrenzeit und nicht auf allophatische Beimengungen in Gestalt von Tephren zurückführte (SCHÖNHALS 1957a: 10ff.; SCHÖNHALS 1957b: 385; BRUNNACKER 1965: 65; MAHR 1996: 21). Erste Beschreibungen eines als typische Lockerbraunerde bezeichneten Bodentyps mit einer dunkel gefärbten sowie hohlraumreichen Hauptlage, welche sich innerhalb äolisch transportierter und teilweise solifluidal akkumulierter Vulkanasche entwickelte, wurden deshalb erstmals durch STÖHR im Jahre 1963 für den Bereich des Hunsrück vorgenommen (STÖHR 1963: 332f.).

BRUNNACKER diskutiert in den Folgejahren weiter die pedogenetischen Prozesse, welche nicht-allophatische Lockerbraunerden entstehen ließen. Dabei kann er jedoch, zumindest für den von ihm untersuchten Bereich des Bayerischen Waldes, das von BARGON fünf Jahre zuvor beschriebene, solifluidal aufgearbeitete Substrat tertiärer Böden oder einen Podsol mit einem denudiertem Eluvialhorizont nur schwerlich nachvollziehen (BARGON 1960: 229; BRUNNACKER 1965: 65). Eine Entstehungsmöglichkeit für diesen insbesondere in nördlicher und östlicher Exposition sowie in Höhenlagen von etwa 800 – 1150 m NHN auf kristallinem Gestein existenten Bodentypen sieht BRUNNACKER in der solifluidalen Aufarbeitung des sesquioxidreichen Unterbodens von Podsolen, welcher infolge einer schützenden Firnschicht keiner folgenden Denudation unterlag (BRUNNACKER 1965: 66ff.). Plausibler scheint ihm jedoch eine rezente Genese der nicht-allophatischen Lockerbraunerde auf Grundlage stetiger Durchfeuchtung des Oberbodens infolge hoher Niederschläge sowie hoher Luftfeuchtigkeit, welche unterstützt durch einen hohen Anteil an organischer Substanz rege Wechsel zwischen Oxidations- und Reduktionsprozessen erwirkt, wobei dieser Umsatz an Metallverbindungen stetige Änderungen der Volumenverhältnisse und eine erhöhte mechanische Verwitterung mit Erhöhung des relativen Anteils an Schluff mit sich brachte, sodass bodenchemische also letztlich bodenphysikalische Veränderungen hervorriefen (BRUNNACKER 1965: 74f.).

Die beiden Autoren MEYER und SAKR publizieren im Jahre 1970 weiterhin einige Veröffentlichungen, innerhalb welcher sie sich der Lockerbraunerdegenese aus verwittertem Basalt im Vogelsberg sowie aus Laacher-See-Tephra innerhalb des Westerwaldes sowie des Taunus annahmen (vgl. SAKR & MEYER 1970: 3f.). Sie identifizieren die vulkanischen Gläser, welche leicht zu Allophanen verwittern und sodann die Reaktionen der Teilchenoberflächen dominieren, als wesentliche Elemente der Lockerbraunerdegenese (MEYER & SAKR 1970a: 57ff.). Auf diese lassen sich nach Ansicht der beiden Autoren auch die hohen Werte insbesondere des oxalatlöslichen Eisens wie auch des Aluminiums zurückführen, welche infolge ihrer schlechten Kristallisation wiederum hohe Anteile von organischer Substanz adsorbieren können (MEYER & SAKR 1970a: 57ff.; SAKR & MEYER 1970: 44f.; MAHR 1998: 22). In der darauf folgenden Publikation wird daneben auf die hohe Dispergierungs-Resistenz sowie das thixotrophe Verhalten des Substrates bei der Geländeaufnahme hingewiesen, welches in den untersuchten Bodenprofilen weniger infolge des hohen Gehaltes an organischer Substanz und den schlecht kristallisierten Eisenoxiden auftritt als vielmehr wiederum auf die Allophane und deren regelhaft angeordnetes Plättchengefüge zurückzuführen ist (MEYER & SAKR 1970b: 103).

Im Vogelsberg untersuchte desweiteren AKINCI in Vorarbeit zu seiner Publikation aus dem Jahre 1973 Lockerbraunerden, welche aus Verwitterungsprodukten von Basalttuff und Lösslehm hervorgegangen sind und verglich diese mit sauren Braunerden von Standorten, deren Substrat sich im Wesentlichen aus Verwitterungsprodukten des reinen Basalts sowie des Lösslehms zusammensetzt (AKINCI 1973: 5ff.). Dabei stellt AKINCI neben den hohen Allophangehalten der verwitterten vulkanischen Gläser auch höhere Gehalte an organischer Substanz in den Lockerbraunerden fest, welche er neben der relativ geringen biologischen Aktivität auf die Bewahrung vor einer Remineralisierung durch die Komplexbildung mit Allophanen wie auch mit Sesquioxiden zurückführt (AKINCI 1973: 144f.). Ebenso zeigen sich bei den Lockerbraunerden in Relation zu den sauren Braunerden hohe Gehalte der durch sesquentielle Extraktion quantifizierbaren Aluminium-ionen, welche ebenso wie die enthaltenen Eisenionen zumeist in organischen Komplexen gebunden vorliegen (AKINCI 1973: 158ff.).

Weiterhin beschäftigte sich auch ABO-RADY in den 1980er Jahren mit Lockerbraunerden, welche im hessischen Raum vorzufinden sind, und deren Genese sich auf die Existenz von Basalttuff sowie anteilig von Tephra zurückführen lässt (ABO-RADY 1985: 231ff.). Er verweist bei den von seiner Seite untersuchten Bodenprofilen, deren Oberboden er zumeist als Kolluvium anspricht, insbesondere auf die hohe Konzentration an Schwermetallen innerhalb des Lockerbraunerde-profils, welche infolge eines Auftretens von Eisenoxiden und durch die Komplexbildung mit der organischen Substanz resultieren (ABO-RADY 1985: 246f.).

Ab dem Jahre 1991 greift VÖLKEL erneut das Problem der nicht-allophatischen Lockerbraunerden im Bayerischen Wald auf und stellt bei der Untersuchung der pedogenen Oxide innerhalb des Substrats dieses Bodentyps fest, dass insbesondere der oxalatlösliche Anteil des Eisens im Verwitterungshorizont signifikant erhöhte Werte im Gegensatz zu den Braunerden aufweist (VÖLKEL 1991: 878; VÖLKEL 1995: 117f.). Als Ursache dieser werden durch Völkel prähclozäne Vor-verwitterungsprozesse des Ausgangssubstrats sowie eine mögliche Zufuhr des Eisens durch Lateralfluss diskutiert, wie sie insbesondere von STAHR seit 1979 für den Naturraum des Schwarz-waldes im Sinne der Ockererden diskutiert werden (STAHR 1979: 61f.; JAHN ET AL. 1994: 133ff.; VÖLKEL 1995: 122f.).

MAHR befasst sich in ihrer Publikation aus dem Jahre 1998 ebenso mit der Entstehung der flächenhaft auftretenden, nicht-allophatischen Lockerbraunerden im Bayerischen Wald und führt diese dabei wie schon BRUNNACKER auf schlecht kristallisierte Metallverbindungen zurück, wobei für die lockere Lagerung und das stabile Gefüge innerhalb dieses Untersuchungsraumes ihrer Ansicht nach insbesondere der hohe Anteil an Ferrihydritmineralen zu beachten ist, sodass sie entsprech-end ihrer Befunde eine Abgrenzung dieser Lockerbraunerden von den vorherrschenden Norm-braunerden bei einem Wert von 1 % am oxalatlöslichen Eisengehalt vornimmt (MAHR 1998: 270f.;

MAHR & VÖLKEL 1999: 472f.). Eine Erklärung für die hohen Anteile der schlecht kristallisierten Metall-oxide sieht sie zum Einen in einer optisch nicht wahrnehmbaren "Podsolierung im weitesten Sinne" (MAHR 1998: 272), welche jedoch durch den ebenfalls hohen Anteil an organischer Substanz ver-schleiert wird. Doch hält sie ebenfalls ein Genese der heutigen Lockerbraunerden innerhalb des Substrates einer mesozoisch-teritären Verwitterungsdecke gemäß BARGON für plausibel (MAHR 1998: 276). Denkbar scheint ihr innerhalb ihres zur Erklärung herangezogenen, vielschichtigen Faktorenkomplexes auch eine kleinräumige standörtliche Variation von hydrologischen wie auch klimatischen Faktoren, welche die mikrobielle Aktivität des Bodens maßgeblich beeinflussen, um insbesondere für das Nebeneinander von Lockerbraunerden und Normbraunerden auf identischen Hangpositionen und auf selben Höhenniveau eine Erklärung zu liefern (MAHR 1998: 278; MAHR & VÖLKEL 1999: 475). Eine laterale Zufuhr der Metalle mit dem Hangwasser sieht sie entgegen den Feststellungen, welche von diversen Autoren für den Naturraum des Schwarzwaldes postuliert wurden, jedoch aufgrund des großflächigen Vorkommens der Lockerbraunerde in den Hochlagen des Bayerischen Waldes nicht gegeben, da sich infolge hoher hydraulischen Leitfähigkeit des rela-tiv grobkörnigen Substrates kein gesättigter Hangwasserfluss ergeben würde (MAHR 1998: 274).

2 Naturraumanalyse

2.1 Lage und naturräumliche Einordnung

Der Schwarzwald, welcher sich östlich des Rheins an der deutsch-französischen Grenze auf dem südwestlichen Gebiet des Bundeslandes Baden-Württemberg erstreckt, bildet mit seiner Fläche von etwa 6000 km² eines der größten Gebirge der Bundesrepublik (WILMANNS 2001: 13). Seine Aus-dehnung beträgt von Südsüdwest nach Nordnordost 166 km, wobei über diesen Verlauf hinweg die Ost-West-Ausdehnung dieses Höhenzuges von 30 bis zu einer Breite von 60 km variiert (WILMANNS 2001: 13).

Das gesamte Gebiet, in welchem sich der Schwarzwald befindet, kann der naturräumlichen Großeinheit des Schichtstufenlandes beiderseits des Oberrheingrabens zugeschrieben werden (WILMANNS 2001: 16f.). An seinen westlichen Ausläufern wird der Schwarzwald durch eine Randver-werfung vom Oberrheinischen Tiefland abgegrenzt, an seinen östlichen Ausläufern schließt sich die Region der Neckar- & Täuber-Gäuplatten sowie an seiner südlichen Begrenzung das Hoch-rheingebiet an (MEYNEN ET AL. 1962: 241ff.; WILMANNS 2001: 16f.).

Im Süden gehören dem Gebirge selbst die naturräumlichen Regionen des Hochschwarz-waldes wie auch des Südöstlichen Schwarzwaldes an, im Nordteil ist der Grindenschwarzwald mit den Enzhöhen, zu welchen der Mittlere Schwarzwald das Bindeglied darstellt, von den Schwarz-waldrandplatten und dem Nördlichen Talschwarzwald im Nordosten umrandet (MEYNEN ET AL. 1962: 244; WILMANNS 2001: 14f.).

2.2 Geologie und Relief

Die Gesteine, welche das prävariscische Grundgebirge des Schwarzwaldes ausbilden, erfuhren bereits seit dem Paläozoikum eine polymetamorphe Umwandlung, wobei die Paragneise, welche aus dem Abtragungsschutt präkambrischer Gebirge entstanden sind, das ältest bekannte Gestein des Schwarzwaldes bilden (METZ 1977: 13; WILMANNS 2001: 25). In diesen hauptsächlich aus Grauwacken, Sandschiefern und Arkosen hervorgegangenen und vermutlich im Algonkium bereits in Tiefen von etwa 15 bis 40 km Tiefe metamorphisierten Sedimentserien drangen innerhalb dieses Zeitalters magmatische Intrusivmassen mit primär granitischer Zusammensetzung empor, welche die Gneise über eine Kontaktmetamorphose veränderten oder durch Intrusion in die geschieferten Bereiche Migmatite ausbildeten (METZ 1977: 13; WILMANNS 2001: 27). In höhergelegenen Bereichen der Erdkruste ereignete sich in diesem Zeitraum zwischen Eruptivmassen und Nebengestein die Genese von schollenartigen Gesteinsverbänden (METZ 1977: 13).

Dieser gesamte Gesteinskomplex erfuhr als Bestandteil des Moldanubikums vor etwa 620 Mio. Jahren schließlich eine mechanische Überprägung im Zuge der assynthischen Gebirgsbildung, woraus im Untersuchungsgebiet in größerem Umfang rezent lediglich am Omerskopf oberflächlich anstehende Para- und Orthogneise hervorgingen, welche erst während der variscischen Orogenese und damit einhergehenden magmatischen Vorgängen erneut einer Umwandlung durch die anatektische Aufschmelzung unterlagen (METZ 1977: 13ff.; GÜNTHER 2010: 33f.). In dieser Gebirgsbildungsphase, welche sich ab dem Oberdevon in hauptsächlich vier Phasen vollzog, bildeten sich auf tieferem Krustenniveau granitische Magmatite, die infolge einer intrusiven Migration in fremden Gesteinsrahmen begannen auszukristallisieren, wobei lediglich spätorogene Magmatite mit ausgeprägten granittektonischen Charakteristika im Nordschwarzwald vorzufinden sind (REGELMANN 1935: 32; METZ 1977: 19ff.). Insbesondere die Kalifeldspat-Großkristalle, welche mehrere Dezimeter in ihrer Ausdehnung aufweisen können, bilden das Hauptmerkmal der regionaltypischen Granite (METZ 1977: 21; WILMANNS 2001: 27). Gründe für diese Ausprägung stellen allgemein die langsame Abkühlung der Gesteinsschmelze sowie insbesondere die Kalifeldspatporphyroblastese dar, bei der im Spätstadium der Abkühlung eine Alkalizufuhr in undeformierte Granite erfolgte und entsprechend dieser Piezokristallisation für ein weiteres Wachstum der Kristalle in dem bereits verfestigendem Gestein sorgte (METZ 1977: 21; WILMANNS 2001: 27). Neben den im Nördlichen Talschwarzwald anstehenden Biotitgraniten sind innerhalb des Untersuchungsraumes, wie auf Abbildung 1 auf der nachfolgenden Seite ersichtlich, insbesondere im Grindenschwarzwald Zweiglimmergranite vertreten (METZ 1977: 25; GÜNTHER 2010: 40).

In der Epoche des Oberkarbon entstanden mit dem Ende des variscischen Magmatismus eine Vielzahl an Erz- wie auch Mineralgängen, welche aus hydrothermalen Lagerstätten hervorgegangen sind, wobei insbesondere in der Region des Nordschwarzwalds im Bühlertal- und Forbachgranit kurzstreichende, an Eisenerz reiche, Gänge bekannt sind (METZ 1977: 29f.).

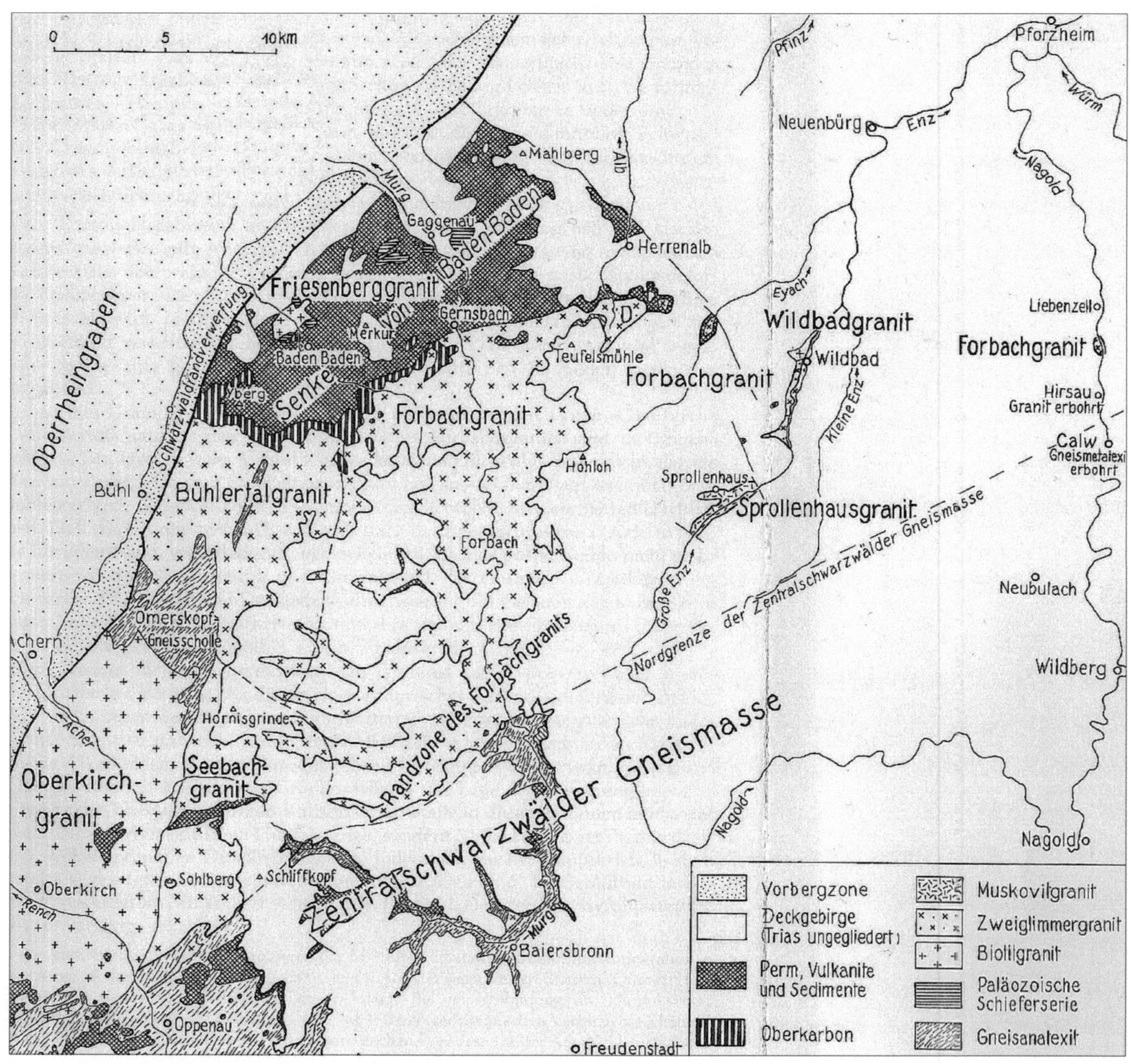

Abb. 1: Das Grundgebirge des Nordschwarzwaldes (verändert nach Metz 1977: 20).

Ebenso ereignete sich in diesem geologischen Zeitabschnitt die Einrumpfung des Höhenzuges, wobei sich in tieferen Reliefpositionen klastische Sedimente und Abtragungsschutt akkumulierten, bevor mit dem Beginn des Rotliegend eine Bruchtektonik einsetzte, welche eine Zerblockung der Gebirgsrumpfflächen und damit einhergehend eine ausgeprägte vulkanische Aktivität herbeiführte, der auch die Förderung rhyolitischer Eruptivmassen in Form von Tuffen und Ignimbriten zuzuordnen ist (Metz 1977: 30f.; Wilmanns 2001: 25). Mit dem Ende des Perms erschien der Schwarzwald infolge dieser Entwicklungen schließlich in Gestalt einer permischen Abrasionsfläche (Regelmann 1934: 20; Metz 1977: 33).

Diesem Grundgebirgssockel auflagernd kann man in den Hochlagen primär postvariscisch aufgelagerte Sedientgesteine des Buntsandsteins vorfinden, welche im Nordschwarzwald größere

Mächtigkeiten als im Südschwarzwald infolge der zeitlichen Variabilität der Ablagerung wie auch der unterschiedlichen Senkung der Teilbereiche des Höhenzuges einnehmen und dem entsprechend eine markante Stufe am Übergang ausbilden, welche speziell im Nordteil des Gebirges ersichtlich ist (METZ 1977: 33; SEMMEL 1996: 139; WILMANNS 2001: 30). Das sandige wie periodisch auch konglomeratische Substrat, aus welchem die Schüttungen bestanden, unterlag dabei mehrfachen Umlagerungsprozessen in ihrem Abtragungs- wie auch dem späteren Sedimentationsgebiet aufgrund von epirogenen Bewegungen (METZ 1955: 34). Der im Nordschwarzwald im Mittel 50 m mächtige Untere Buntsandstein zeigt sich in einer Folge von roten, violetten sowie weißen und gelblichen Sandsteinen, welche schwach verkieselt wie auch teilweise glimmerführend und mit tonigen Einlagerungen wie frischen oder bereits kaolinitisierten Feldspäten versehen sind und dunkelbraune bis schwarze Konkretionen aufgrund von Eisen-Mangan-Oxiden und insbesondere deren Verwitterungsrückständen aufweisen (METZ 1977: 35f.; GÜNTHER 2010: 64). Der Mittlere Buntsandstein setzt sich aus dem geröllführendem Eckschen Konglomerat im Liegenden sowie Hauptkonglomerat im Hangenden zusammen, welchem der geröllfreie Bausandstein von großer Mächtigkeit zwischengeschaltet ist (METZ 1977: 36; GÜNTHER 2010: 72). Das rötlich bis violette Ecksche Konglomerat führt in seiner ebenfalls 50 m mächtigen Schichtfolge neben Quarzen vor allem auch granitische Gesteine und zu einem geringen Teil schwarze Kieselschiefer als Geröll (METZ 1977: 36ff.; WILMANNS 2001: 31). Über dem zwischen 100 und 200 m Mächtigkeit aufweisenden Bausandstein mit hellroter bis bräunlichroter Farbgebung folgt das im Mittel 60 m mächtige Hauptkonglomerat, dessen Geröll sich fast ausschließlich aus den lokal als "Gaggerle" benannten Quarzen und minimal aus glimmer- und feldspatreichen Gesteinen zusammensetzt (METZ 1977: 36ff.; WILMANNS 2001: 31). Der Obere Buntsandstein setzt sich wie in Abbildung 2 ersichtlich dem entgegen innerhalb seiner 50 m Mächtigkeit aus einem geringmächtigen und violettfarbigem Karneolhorizont, welcher von glimmerreichem Plattensandstein und abschließend den an wenigen Stellen erhaltenen Rötton mit einer roten sowie teils grünen Färbung überlagert ist, zusammen (METZ 1977: 39; WILMANNS 2001: 30).

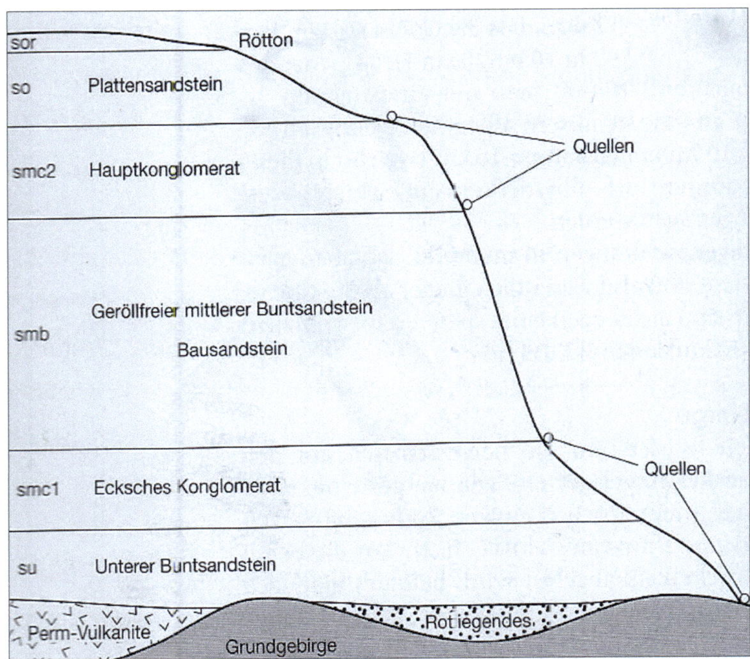

Abb. 2: Das Buntsandsteindeckgebirge mit Quellhorizonten (verändert nach WILMANNS 2001: 30).

In der Zeit des Jungtertiär erfolgte eine ungleichmäßige Emporhebung der einzelnen Bereiche des Höhenzuges, wobei der Feldberg und folgend die Hornisgrinde die höchstgelegenen Schollen bilden, welche von der senkrecht zur Gebirgsausdehnung verlaufenden Kinzigmulde getrennt werden, deren Buntsandsteinauflagerung mehrere hundert Meter niedriger vorzufinden ist (Metz 1977: 41). Infolge dieses Hebungsprozesses entstand die Schwarzwaldrandverwerfung sowie eine Vorbergzone aus zumeist antithetisch zerblockten Schollen, die dem Schwarzwald über seinen Verlauf hinweg westlich vorgelagert ist (Metz 1977: 41ff.). In Gangräumen, welche sich infolge dieser tektonischen Tätigkeit im Deckgebirge ergaben, bildeten sich hydatogene Mineralgänge aus, die aus einer mit relativ kühler Lösung angereicherten vadosen Zone selektiv durch hydrothermale Erzgänge des Grundgebirges beliefert wurden (Günther 2010: 138ff.; Metz 1977: 42). Im ausgehenden Tertiär bis in das Pleistozän erfolgte zeitgleich zum Hebungsprozess eine Denudation, durch welche der Schwarzwald erneut eine flachreliefierte Landschaft erlangte, welche von breitsohligen Tälern durchzogen war (Metz 1977: 44ff.).

Innerhalb der folgenden Eiszeiten brachte einzig die Würmeiszeit eine ausgedehnte Eisbedeckung auf dem Feldberg wie auch der Hornisgrinde mit sich, deren Gletscherzungen weit in die montane Stufe herab reichten; im Periglazialgebiet der reliefniederen Bereiche unterhalb der Schneegrenze wurde infolge geringen Bedeckungsgrades durch die Vegetation hingegen äolisches Lockergestein vornehmlich alpiner Herkunft akkumuliert (Metz 1977: 46f.; Jahn et al. 1994: 28; Semmel 1996: 135).

2.3 Oberflächennaher Untergrund

Die Böden der höheren Gebirgslagen des Schwarzwaldes sind zumeist aus dem vergrusten Zersatz des Anstehenden sowie seltener aus groben Bruchstücken des Ausgangsgesteins hervorgegangen, welchem in den Hangbereichen die Basislage folgt, während die darauf auflagernde Hauptlage in ubiquitärer Verbreitung vorhanden ist (Semmel 1993: 47; Jahn et al. 1994: 28). Daneben zeigt sich in den Hochlagen an Steilhängen sowie an Gesteinsausbissen eine Erscheinung, die als Oberlage interpretiert werden kann (Semmel 1993: 47). Punktuell treten zudem vornehmlich im Bereich des Grundgebirges an der Oberfläche auch Felsenmeere und Blockhalden zutage, welche infolge von Wollsackverwitterung und nicht dem Nachbrechen von Gesteinsblöcken, wie es im Bereich des Deckgebirges ersichtlich ist, entstanden sind (Metz 1977: 66f.; Wilmanns 2001: 200). In Bereichen der würmeiszeitlichen Vergletscherung auf der Hornisgrinde sowie an den ostexponierten Hängen der Muldentäler ist der Untergrund hingegen aus Geschiebelehm aufgebaut, der von einer geringmächtigen Hauptlage überdeckt ist (Jahn et al. 1994: 28).

Innerhalb der Hauptlage sind in Bereichen des ehemaligen Periglazialgebietes relativ geringe Mengen an Lösslehm, welcher aus Fernlöss oder aber auch lokalen Einwehungen generiert worden sein kann, wie auch geringe Anteile der Tephra des Laacher-See-Vulkans vorzufinden (Metz 1977: 46f.; Jahn et al. 1994: 28; Schmincke 2000: 171).

Neben den teilweise im Löss entwickelten Übergangsböden von Parabraunerden zu Braunerden an den Unterhängen des Höhenzuges zeigt in den höheren Lagen des Grundgebirgsschwarzwaldes neben lokaler Vorkommen von Rankern in diversen Varietäten und Syrosemen generell die Braunerde in verschiedenen Ausprägungen die maßgebende Verbreitung; neben Sauer- und Humusbraunerden nimmt dabei insbesondere die mittelgründige Normbraunerde, teilweise mit initialen bis geringfügigen Podsolierungserscheinungen, eine vorrangige Stellung ein (RADKE 1973: 18f.; JAHN 1994: 28f.; AG BODEN 2005: 215f.).

In den steilhangigen Tälern sowie an den häufig vorzufindenden Quellaustritten am Übergang vom Grund- zum Deckgebirge sind desweiteren lokal beschränkte Subtypen des Gleys vom Quell- über den Anmoor- bis hin zum Hanggley vertreten; ebenere Bereiche können ebenso Stagnogleye und Niedermoore aufweisen (JAHN 1994: 29).

Auf den Hängen des Deckgebirges sowie aus dessen Gestein hervorgegangenen periglazialen Lagen hangabwärts sind hingegen lediglich podsolige Braunerden sowie zum Großteil Podsole in verschiedenen Subtypen verbreitet, da das grobkörnige, feinmaterialarme Ausgangssubstrat und der hohe Niederschlag in Kombination den Prozess der Podsolierung begünstigen (RADKE 1973: 18f.). Auf den Hochflächen des Schwarzwaldes hingegen sorgen schießlich entsprechende geologische Verhältnisse aus lehmigem bis tonigem Substrat für einen Wasserstau, sodass sich auf diesen großflächig Stagnogleye und Hochmoore etablieren konnten (RADKE 1973: 26f.; METZ 1977: 74; JAHN 1994: 29).

2.4 Klima und Hydrologie

Der Schwarzwald stellt eines der am stärksten durch ozeanische Luftströmungen geprägten Mittelgebirge der Bundesrepublik dar, was durch verhältnismäßig milde Winter wie auch die höchsten Niederschlagswerte im außeralpinen Bereich zum Ausdruck kommt (METZ 1977: 70; WILMANNS 2001: 18). Dies resultiert aus dem Umstand, dass einzig der vorgelagerte Höhenzug der Vogesen die niederschlagsreichen Westwinde vom Atlantik abmildert (METZ 1977: 70; WILMANNS 2001: 18). Wie auf Abbildung 3 auf der nachfolgenden Seite ersichtlich, reichen die Niederschlagssummen im jährlichen Mittel von etwa 1000 mm/a an den Ausläufern dieses Gebirges bis hin zu 2000 mm/a an Kuppen im Nordschwarzwald wie etwa der Hornisgrinde, da westlich dessen die stauende Wirkung der Hochvogesen entfällt und so eine Aufgleitfront entsteht (METZ 1977: 70; WILMANNS 2001: 19ff.). In Höhenlagen über 900 m zeichnet sich dabei neben dem Niederschlagsmaximum in den Sommermonaten ein zweites im Januar eines jeden Jahres ab (WILMANNS 2001: 21).

Die jährlichen Temperaturmittel bewegen sich in einem Bereich von etwa 9 bis 10 °C am Westrand des Schwarzwaldes, wobei sie mit der Höhe auf einen Wert zwischen 6 und 7 °C in den hohen Mittelgebirgslagen um etwa 900 m sinken, bevor sie auf der Hornisgrinde mit 4,9 °C und auf dem Feldberg mit 3,2 °C die niedrigsten Werte im gesamten Schwarzwald erreichen (METZ 1977: 69; HÖLZER & HÖLZER 1995: 201; WILMANNS 2001: 22).

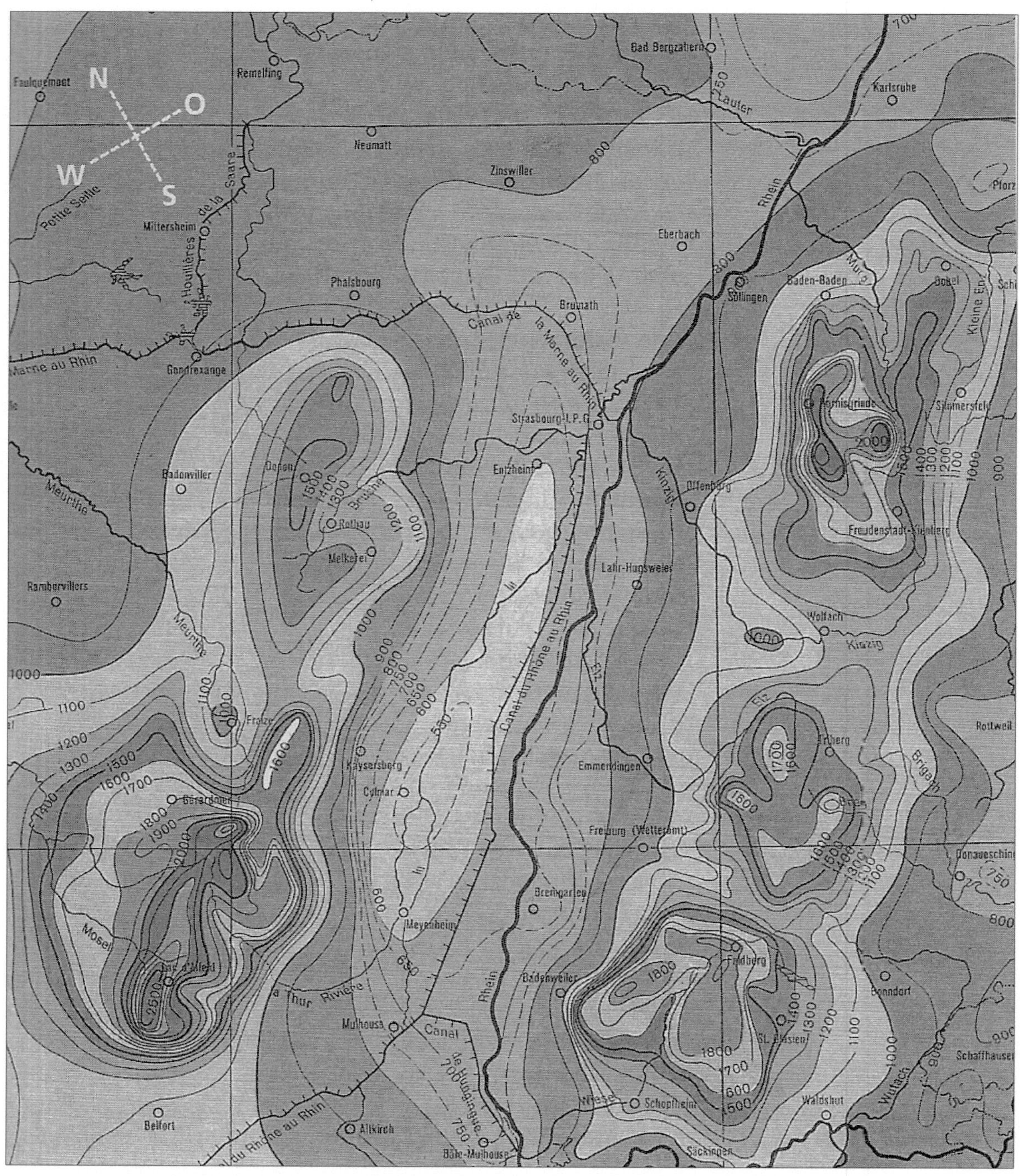

Abb. 3: Jährliche Mittel der Niederschlagssummen im Bereich Vogesen und Schwarzwald (verändert nach WILMANNS 2001: 19).

Kaltluftseen infolge von Inversion bilden sich insbesondere bei starker Ausstrahlung und windstillen Verhältnissen in den durch steile Hänge gesäumten Taleinschnitten aus; die Ausbildung einer nebelreichen Sperrschicht wie auch die Herausbildung von Kammeis in den Böden können sich zudem daraus ergeben (WILMANNS 2001: 22).

In Synthese von Niederschlag und Relief hat sich am Höhenzug des Schwarzwaldes die europäische Wasserscheide ausgebildet, welche das danubische wie auch das rhenanische Gewässersystem durch seine zahlreichen Bach- und Flussläufe speist, wobei Letzteres ein Wachstum infolge erhöhter Erosionstätigkeit seiner Gewässerläufe seit dem Jungtertiär verzeichnen kann (WILMANNS 2001: 33). Das gesamte Gewässersystem des Nordschwarzwaldes gehört dabei zum Stromgebiet des Oberrheins, da die Gewässerläufe entweder direkt in diesen entwässern oder ihr Durchfluss über den Neckar zugeführt wird (METZ 1977: 68). Ihren Ursprung nehmen diese Fließgewässer insbesondere in den zahlreichen Rheokrenen, welche sich zum Hauptteil an der Untergrenze des Auflagerungshorizontes des Buntsandsteins wie auch oberhalb des Eckschen Konglomerats oder an den Grenzen des Hauptkonglomerats infolge wasserstauender Eigenschaften ausgebildet haben (METZ 1977: 69; WILMANNS 2001: 30, 48).

Stehende Gewässer haben sich punktuell in Vielzahl vor allem als Karseen in Hohlformen der letzten Eiszeit im Deckgebirge ausgebildet, welche insbesondere den Bereich um den Feldberg großflächig vergletscherte und im Nordschwarzwald lediglich die höchsten Lagen kleinräumig mit Firnschnee bedeckte (WILMANNS 2001: 34). Ebenso bilden die Hochmoorflächen auf den Bergkuppen des Höhenzuges wie auch punktuell in den Karen stehende Gewässer, deren Existenz auf die hohen Niederschlagsmengen sowie ebenfalls auf die stauenden Eigenschaften der geologischen Schichten zurückzuführen ist (METZ 1977: 74f.; WILMANNS 2001: 49ff.).

2.5 Vegetation

Der Höhenzug des Schwarzwaldes, der das größte Waldgebiet der Bundesrepublik darstellt, zeichnet sich rezent im Wesentlichen durch Nadelholz aus, welches im Zuge der planmäßigen Forstwirtschaft des 18. Jh. in vielen Wäldern der Mittelgebirge in Ausprägung der Fichte vermehrt Einzug hielt, obgleich in den natürlichen Waldgesellschaften auch der Laubbaumbestand von Relevanz ist (HASEL 1944: 76; METZ 1977: 78).

Die natürlichen Waldgesellschaften des Schwarzwaldes bilden indes Hainsimsen-Buchenwälder mit stellenweisem Vorkommen der Trauben-Eiche und der Ess-Kastanie an nährstoffarmen sowie Waldmeister-Eichen-Buchenwald an nährstoffreichen Standorten der submontanen Stufe (RADKE 1973: 21; WILMANNS 2001: 70f.). In der montanen Stufe bildet das Vorkommen der Tanne ein wesentliches Element, wobei speziell nährstoffreiche Standorte innerhalb dieses Bereichs einen Waldschwingel-Tannen-Buchenwald ausbilden, welcher in der hochmontanen Stufe eine Entwicklung hin zum Bergahorn-Buchenmischwald vollzieht, während an den weit verbreiteten, nährstoffärmeren Standorten lediglich die Fichte einen natürlichen Bestand im bereits bestehenden Hainsimsen-Buchenwald ausbildet (HASEL 1944: 77; WILMANNS 2001: 70f.). An den Hängen des danubischen Gewässersystems ist natürlicherweise hingegen Beerstrauch-Kiefern-Tannenmischwald ausgebildet, der mit Labkraut-Tannenwald in allen Höhenstufen korrespondiert, wobei diese Bereiche aufgrund ausgeglicheneren Reliefs der Fichtenbepflanzung unterliegen (WILMANNS 2001 71f.).

Wald-Sonderstandorte sind dem entgegen etwa an Bach- und Flussläufen mit Auwäldern eines Schwarzerle-Uferwaldes oder eines Bach-Eschenwaldes sowie die Schluchtwälder, welche sich über einen Ahorn-Lindenwald mit ebenso vereinzelten Exemplaren von Esche und Berg-Ulme zusammensetzen (RADKE 1973: 29; WILMANNS 2001: 74f.). Weitere Bereiche, innerhalb welcher die natürlichen Vorkommen der Tanne und Buche keinen Standortvorteil verzeichnen können, ergeben sich etwa auf flachgründigen Böden der weit verbreiteten Blockmeere und auf Felsvorsprüngen, wo die Hänge-Birke sowie die Vogelbeere verbreitet ist oder aber bei besonders nährstoffarmen Verhältnissen ein natürlicher Moos-Kiefernwald ausgebildet sein kann, wie dies auch in den Rand-bereichen der Moorkiefernwälder in der hochmontanen Stufe ersichtlich ist (RADKE 1973: 30; WIL-MANNS 2001: 74).

Neben diesem Großteil an bestockter Fläche finden sich auch ausgedehnte Areale, die durch Borstgrasrasen in den montanen Bereichen das Kulturgrünland repräsentieren (WILMANNS 2001: 55). An Bachläufen oder quelligen Bereichen, welche dieses durchziehen, können die Waldsimsenwiese, die Mädesüß- wie auch die Quellstaudenflur ausgebildet sein (WILMANNS 2001: 59). In niedrig gelegeneren Reliefpositionen sind neben wenigen Mager- vor allem Fettwiesen in der Ausbildung von Glatthaferwiesen wie auch Goldhaferwiesen vorzufinden (WILMANNS 2001: 59).

Auf den Grinden haben sich dem entgegen teilweise Hochmoore wie auch Missen aus-gebildet, die bis auf einige Exemplare der Moorkiefer waldfrei und aus diversen Torfmoosen der Gattung Sphagnum oder zumeist als Rasenbinsenmoore ohne jegliche Torfmoose ausgebildet sind (RADKE 1973: 30). Weitere Nassbereiche werden oftmals von einer torfmoosreichen Rasenbinsen-Gesellschaft umsäumt, die in den Hochlagen in die Bergheidegesellschaft übergeht, welche typischerweise als Krautschicht der Fichtenforste ausgebildet ist (RADKE 1973: 30; HÖLZER & HÖLZER 1995: 208; WILMANNS 2001: 75). Aufgrund der hohen Niederschläge und der geologisch bedingten Quellbereiche nimmt zudem die Vegetation der Quellfluren eine besondere Stellung ein, die im Schwarzwald primär in Form des Gegenblättrigen Milzkrauts in Erscheinung tritt (WILMANNS 2001: 48). Bezeichnende Vorkommen sind jedoch auch von der Quell-Sternschmiere, dem Geschlängel-tem Schaumkraut und dem Bitter-Schaumkraut zu verzeichnen, zu welchem sich in der subalpinen Stufe noch Quellflur-Weideröschen-Arten und der Stern-Steinbrech gesellen (WILMANNS 2001: 48).

2.6 Nutzungshistorie

Erste Besiedlungsbestrebungen des Nordschwarzwaldes gab es bereits vor etwa 2.500 Jahren durch die Kelten im Raum von Neuenbürg (GÜNTHER 2010: 159). In der weiteren Umgebung kann durch kulturhistorische Landschaftselemente wie Schachtöfen der Frühlatènezeit bereits für diesen Zeitraum auf eine industrielle Nutzung des Schwarzwaldes im Zuge der Eisenerzverhüttung geschlossen werden; erst im Mittelalter setzte zudem der Abbau von Silbererz sowie dessen Gewinnung über das Seigerhüttenverfahren ein (GASSMANN 2005: 61; GÜNTHER 2010: 161). Um das 15. Jh. begann ebenso die Besiedlung der Hochlagen infolge der Errichtung von Glashütten; auf eine solche geht auch die Gründung der nördlich der Hornisgrinde im Untersuchungsgebiet befindlichen Siedlung Herrenwies zurück (GÜNTHER 2010: 161). Ortsbezeichnungen wie "Aschenplatz" oder "Dreikohlplatten" zwischen Hundseck und Hundsbach im selben Areal zeigen zudem reliktische Nutzungen des gesamten Bereichs durch die mit der Glasherstellung einhergehende Köhlerei an und geben einen Hinweis auf großräumige Entwaldungen (vgl. TK L 7314 BADEN-BADEN).

Rezent zeichnen sich die Hochlagen des Nordschwarzwaldes zu einem Großteil wieder durch Nadelwaldbewuchs aus, welcher die Grundlage der touristischen Erschließung der Region darstellt. Nahe den kleinräumigen Ansiedlungen findet sich eine Vielzahl abgelegener Kurhäuser; als verbindende Elemente durch die Waldareale wurden neben befestigten Fahrbahnen auch Loipen und Wanderwege angelegt (vgl. TK L 7314 BADEN-BADEN). Sondernutzungen auf größerer Fläche stellen desweiteren der SWR-Sendeturm auf der Hornisgrinde, die Schwarzenbachtalsperre östlich Herrenwies sowie die ausgewiesenen Naturschutzgebiete der Hochmoore auf einem Großteil der Bergkuppen dar (vgl. TK L 7314 BADEN-BADEN).

3 Methoden

3.1 Geländearbeit

Bei einer ersten Übersichtsbegehung im November des Jahres 2012 wurde auf Grundlage der Bodenkundlichen Karte BÜK CC 7910 Freiburg-Nord ein grober Einblick die vorherrschenden geologischen Verhältnisse sowie ein Überblick über vorzufindende Bodentypen im Gebiet des Nordschwarzwaldes um die Hornisgrinde erworben. Dabei zeigten sich lediglich die Hochlagen des granitischen Grundgebirges von Relevanz, da innerhalb dieses Bereichs signifikant locker gelagerte Hauptlagen mit einem ungewöhnlich hohem 'greasing-effect" aufgefunden werden konnten, welche für eine anschließende Schwermineralanalyse beprobt wurden.

Im Verlauf einer zweiten Begehung im April 2013 wurden Standorte in diesem Bereich erneut begangen und mit einem Klappspaten an Wegesrändern wie auch Grubenrändern reliktischer Baumwürfe bis höchstens etwa 100 cm Tiefe freigelegt. Hierbei wurde in der näheren Umgebung vermuteter Lockerbraunerdestandorte jeweils ein weiteres Profil ohne bezeichnende Charakteristika für einen Vergleich freigelegt. Auf Grundlage der bodenkundlichen Kartieranleitung in der 5. Auflage aus dem Jahre 2005 wurde folgend eine Profilansprache durchgeführt wobei das "Formblatt für die bodenkundliche Profilaufnahme" (AG BODEN 2005: 46f.) in verkürzter Form Anwendung fand. Die Humusform wurde dabei durch horizontweises Abtragen bestimmt.

Daran anschließend fand eine horizontbezogene Beprobung mit einem Stechzylinder von 250 cm³ Volumen statt, wobei die Probe für den Transport zum Labor in Kunststofftüten überführt wurde, nachdem überstehendes Substrat an den Kopfseiten des Stechzylinders sauber mit einem Messer entfernt wurde.

3.2 Probenaufbereitung

Im Labor fand die Trocknung der Bodenproben im Gefriertrockenschrank Christ Alpha 2-4 LD Plus über einen Zeitraum von 24 Stunden bei -67 °C statt, sodass die Bestimmung des Kohlenstoffgehaltes und der Korngröße an diesen Proben weiterhin möglich blieb. Daran schloss sich die Aufbereitung der Proben für die weitere Analyse durch Mörsern in einem Porzellanmörser sowie eine Trennung des Skelettanteils (> 2 mm) vom Feinboden (< 2 mm) mittels Siebung und anschließender Homogenisierung für die Verwendung in den folgenden Analysen an.

3.3 Laboranalysen

3.3.1 Bestimmung der Trockenrohdichte

Das gefriergetrocknete Bodensubstrat wurde auf einer geeichten Waage für jede Probennummer separat eingewogen. Daran anschließend wurde eine Differenz dieses Trockengewichts und des 250 cm³ umfassenden Gesamtvolumens des verwendeten Stechzylinders gebildet, um die Trockenrohdichte des beprobten Bodenhorizontes zu erhalten.

3.3.2 Bestimmung des Porenvolumens

In einen Messkolben wurden jeweils 10 g des horizontweise entnommenen Bodensubstrats gefüllt, welcher anschließend unter mehrmaligem Schütteln eine Anreicherung mit Methanol (CH_3OH) bis zu einer definierten Marke von 50 ml aus einer Bürette mit einem Startvolumen von 50 ml erfuhr. Eine organische Flüssigkeit wurde bei dem Auffüllversuch dem Wasser (H_2O) vorgezogen, um die Oberflächenspannung dieser Flüssigkeit auszuschließen, wobei selbst organische Festsubstanzen innerhalb der Probe eine Benetzung erfuhren, sodass das gesamte Spektrum der Feinporen erfasst werden konnte.

Die Versuchsreihe wurde aufgrund minimaler bis ausbleibender Abweichungen zweifach durchlaufen, sodass die Streuweite der Werte durch die Verwendung des Mittelwertes minimiert werden konnte. Die Berechnung des Porenvolumens erfolge darauf folgend über die Gleichung 100 % - [(Trockenrohdichte * Volumen der Einwaage) / Einwaage].

3.3.3 Bestimmung des pH-Wertes

Die Ermittlung des pH-Wertes erfolgte mittels elektrometrischer Messung nach DIN 19683, Teil 1 aus dem Jahre 1977 in 0,01 M Kalziumchlorid-Lösung ($CaCl_2$) mit der Einstabmesskette des Typs WTW 740 an zuvor auf 10 g eingewogenem Probenmaterial.

3.3.4 Bestimmung des Kohlenstoffgehaltes sowie der organischen Substanz

Die Messung des Gesamtkohlenstoffgehaltes wurde entsprechend der E DIN ISO 10 694 aus dem Jahre 1994 über die trockene Verbrennung im Sauerstoffstrom mit anschließender Analyse des freigesetzten Kohlendioxids (CO_2) mittels einer Infrarotdetektion am Kohlenstoff-Analysator LECO EC-12 vorgenommen, wobei von dem Substrat wie auch von dem Standard, dessen Verbrennung vor und nach der Messreihe erfolgte, je 3 g eingewogen und mit jeweils einer Spatelspitze Kupfer- wie auch Eisenspäne versehen wurde.

Da sich die pH-Werte der Proben in den sauren Bereich einordnen ließen, war von einer Freisetzung weiteren Kohlendioxids durch Carbonate nicht auszugehen. Eine Berechnung der Gehalte an organischem Kohlenstoff erfolgte anschließend über die Multiplikation der Messergebnisse mit dem Faktor 1,72 (AG Boden 2005: 111).

3.3.5 Bestimmung der Korngröße

Die Bestimmung der Korngröße wurde der DIN 19 683, Teil 1 und 2 aus dem Jahre 1973 entsprechend vorgenommen, wobei aufgrund des zuvor ermittelten hohen Anteils an organischer Substanz eine Humuszerstörung durch Ansetzen der Probe mit Wasserstoffperoxid (H_2O_2) über Nacht sowie einem vierstündigen Kochvorgang am Folgetag durchgeführt werden musste. Anschließend fand die Dispergierung mit 0,4 N Natriumpyrophosphat ($Na_4P_2O_7$) statt, sodass die Analyse der Kornfraktionen bis 63 µm mittels Nasssiebung und jene der Kornfraktionen unter 63 µm entsprechend der Sedimentationsanalyse nach Köhn durchgeführt werden konnte.

3.3.6 Sesquentielle Extraktion von pedogenem Eisen, Aluminium & Mangan

Die Bestimmung des dithionitlöslichen Anteils der Eisen- , Aluminium- und Manganverbindungen wurde nach der Methode von Mehra & Jackson aus dem Jahre 1960 mit einer Citrat-Bicarbonat-Dithionit-Lösung ($C_6H_5Na_3O_4$-$NaHCO_3$-$Na_2S_2O_4$) vorgenommen. Dem entgegen fand die Bestimmung der oxalatlöslichen Anteile der genannten Verbindungen entsprechend der DIN 19684, Teil 6 des Jahres 1977 mit oxalsaurem Ammoniumoxalat ($N_2H_8C_2O_4$) statt.

Die Messungen erfolgten jeweils mit dem Flammen-AAS Perkin Elmer AAnalyst 300, dessen Flamme bei der Analyse der dithionit- wie auch oxalatlöslichen Anteile von pedogenem Eisen und Mangan durch ein Acetylen-Luft-Gemisch sowie bei der Analyse des pedogenen Aluminiums durch ein Acetylen-Lachgas-Gemisch gespeist wurde.

3.4 Potentielle Fehlerquellen

Obgleich alle Arbeitsschritte der qualitativen wie vor allem auch quantitativen Auswertung mit äußerster Sorgfalt vorgenommen wurden, sei an dieser Stelle auf potentielle Quellen von Ungenauigkeiten beziehungsweise Fehlergebnissen verwiesen, welche dennoch auftreten können.

Zum einen stellt die Wahrnehmung des Farbtons sowie die Schätzung einer Häufigkeitsverteilung im Gelände, seien es grusige Bestandteile oder auch Wurzeln, einen relativ subjektiven Prozess dar, der bei mehrmaliger Durchführung zu unterschiedlichen Ergebnissen führen kann.

Desweiteren ergibt sich für die Entnahme von Bodensubstrat mit einem Stechzylinder insbesondere bei einem hohen Anteil an Grus im Bodenhorizont und des relativ lockeren Gefüges eine gewisse Fehleranfälligkeit, da aufgrund dessen die Entnahme einer exakt zylindrischen Probe mit einem geraden Schnitt an den Grundflächen schwer zu vollziehen war. Weiterhin ergeben sich auch durch die Überführung in Transport- und Laborbehältnisse, obgleich immer mit mehrmaliger Spülung vollzogen, geringe Verluste, welche das Ergebnis des Trockenrohgewichts beeinflussen.

Schließlich ergibt sich auch für die Messungen am Flammen-AAS eine Reihe möglicher Störungen und Fehlerquellen, welche die Reproduzierbarkeit der Ergebnisse in negativer Weise beeinflussen. Insbesondere sei hierbei auf spektrale Interferenzen, welche auftreten können, wenn andere als die zu messenden Elemente in gleicher Wellenlänge Licht absorbieren, als auch auf Schwankungen bei der Strömungsintensität des Gases sowie der Brennergeschwindigkeit und dessen Temperatur verwiesen (vgl. MAHR 1998: 69f. nach HEINRICHS ET AL. 1985). Ebenso ergibt sich eine weitere potentielle Fehlerquelle über die Eichkurven, die für jede Messreihe neu angelegt werden müssen. Durch Messungen im unteren Bereich der Kurve wurde versucht, Abweichungen vom Lambert-Beerschen-Gesetz zu umgehen, welche bei höheren Extinktionen im verflachenden Bereich er Kurve auftreten können. Ebenso wurde durch eine zweite Messung des Standards am Ende einer jeden Messreihe die Empfindlichkeit und andere Messbedingungen auf mögliche Variabilität hin überprüft, um Ungenauigkeiten der Messergebnisse nachzuvollziehen und die Messung in einem solchen Fall wiederholen zu können (vgl. MAHR 1998: 70 nach HEINRICHS ET AL. 1985).

4 Ergebnis

4.1 Bodenkundliche Beschreibung der Profile

4.1.1 BC-NSW 13 A

Das Profil BC-NSW 13 A (Zone 32 N: 442330 E; 5389650 N) befindet sich auf etwa 610 m über NHN im Bereich einer kleinen Blockhalde, welche sich entlang eines nordostexponierten Hanges nahe des Gertelsbaches oberhalb der Gertelsbachhütte erstreckt (Anhang 1). Neben Buchen und Eichen prägen insbesondere Fichten die Vegetation des Standortes, der außer vereinzelten Vorkommen von Frauenhaarmoos keine großräumige Bodenbedeckung aufweist.

Abb. 4: Lichtbild des Lockerbraunerdeprofils
BC-NSW 13 A (eigene Aufnahme).

Die Bodenform dieses Standortes wurde als eine Lockerbraunerde (Ah/Bfv/IIilCv) angesprochen, deren LH (Zweiglimmergranit, Löss & LST) über einer LB (Zersatz des Zweiglimmergranit) entwickelt und von einer organischen Auflage im Grundgerüst eines F-Mulls zur Geländeoberkante hin begrenzt ist. Innerhalb der LH sind ein Ah-Horizont mit einer Mächtigkeit von 10 cm und unterhalb folgend ein Bfv-Horizont mit einer solchen von 70 cm wie in Abbildung 4 ersichtlich ausgebildet. Neben einer leichten Farbvariation von ockerbraun zu ocker-hellbraun sind insbesondere die zu erwartende Abnahme der Durchwurzelung von Wg3/Wf3 auf Wg1/Wf0 sowie eine Zunahme des Skelettgehalts von Gr2 auf Gr5 und ein Wechsel von Sl4 auf Ls4 in der Korngrößenverteilung des Feinbodens bei gleichbleibendem Kohärentgefüge mit dem Übergang von der LH zur LB festzustellen (Anhang 2).

Innerhalb der beiden innerhalb der LH entwickelten Bodenhorizonte zeigen die oxalatlöslichen Metallverbindungen einen Anstieg von 0,445 auf 0,790 % für pedogenes Eisen, von 0,005 auf 0,011 % für pedogenes Mangan und vor 0,321 auf 0,619 % für pedogenes Aluminium mit der Tiefe. Die dithionitlöslichen Bestandteile steigen von 0,769 auf 1,265 % für das Eisen, von 0,008 auf 0,016 % für Mangan und jene des Aluminiums von 0,320 auf 0,673 % an (Anhang 3).

4.1.2 BC-NSW 13 B

Das Vergleichsprofil zu dem Profil BC-NSW 13 A, das mit der Kennung BC-NSW 13 B versehen wurde, befindet sich lediglich wenige Meter hangunterhalb dessen auf etwa 600 m über NHN (Zone 32 N: 442320 E; 5389670 N) außerhalb der Blockhalde an einer Wegböschung (Anhang 1). Neben Buchen und Eichen finden sich vornehmlich die forstwirtschaftlich geförderten Fichten.

Die Bodenform an diesem Standort wurde als eine geringfügig podsolige Braunerde (Aeh/Bsv/IIiICv) angesprochen, deren LH (Zweiglimmergranit, Löss & LST) über einer LB (Zersatz des Zweiglimmergranit) entwickelt und von einem F-Mulls an der Oberfläche bedeckt ist. Innerhalb der LH sind wie in Abbildung 5 ersichtlich ein Aeh-Horizont mit einer Mächtigkeit von 7 cm und ein Bsv-Horizont mit einer solchen von 45 cm ausgebildet; unterhalb schließt sich eine teils mit Blöcken des Anstehenden durchsetzte Hauptlage an. Neben einer leichten Farbvariation von gelbbraun zu gelb-hellbraun geht eine Abnahme von der Durchwurzelung von Wg2/Wf1 auf Wg1/Wf0 sowie ein Anstieg des Skelettgehaltes in allen Größenordnungen von Gr2 auf Gr5 bei gleichbleibendem Kohärentgefüge und ähnlicher Korngrößenverteilung innerhalb der Kategorie Ls4 mit dem Übergang von der LH zur LB einher (Anhang 2).

Im Tiefenverlauf innerhalb der LH zeichnet sich bei den oxalatlöslichen Metallverbindungen eine negative Entwicklung von 0,534 auf 0,420 % für das pedogene Eisen ab, das pedogene Mangan hingegen lässt einen Anstieg von 0,034 auf 0,038 % ebenso wie bei dem pedogenen Aluminium von 0,469 auf 0,642 % ersichtlich verzeichnen (Anhang 3).

Abb. 5: Lichtbild des podsoligen Braunerdeprofils
BC-NSW 13 B (eigene Aufnahme).

4.1.3 BC-NSW 13 C

Das Profil BC-NSW 13 C (Zone 32 N: 444390 E; 5387290 N) befindet sich auf etwa 780 m über NHN ebenfalls im Bereich einer der im kristallinen Nordschwarzwald sehr verbreiteten kleinen Blockhalden, welche sich entlang eines südwestexponierten Hanges unterhalb der Landstraße 80b zwischen Hundseck und Hundsbach nahe des Greßbaches erstreckt (Anhang 1). Neben Buchen finden sich insbesondere junge Tannen und Fichten in der Baumschicht an diesem Standort, dessen Krautschicht zudem vornehmlich aus Frauenhaarmoos und Waldschwingel aufgebaut ist.

Abb. 6: Lichtbild des Lockerbraunerdeprofils BC-NSW 13 C (eigene Aufnahme).

Die Bodenform dieses Standortes wurde als eine Lockerbraunerde (Ah/Bfv/IIiICv) angesprochen, deren LH (Zweiglimmergranit, Sandstein des Buntsandsteins, Löss & LST) über einer LB (Zersatz des Zweiglimmergranits) entwickelt ist und von einer organischen Auflage eines F-Mulls bedeckt wird. Wie in Abbildung 6 zu sehen, ist innerhalb der LH ein Ah-Horizont mit einer Mächtigkeit vor 5 cm und ein unterhalb folgender Bfv-Horizont mit einer solchen von 35 cm entwickelt. Neben einem Farbwechsel von ockerbraun zu dunkelbraun sind die Abnahme der Durchwurzelung von Wg2/Wf2 auf Wg0/Wf0 sowie eine Zunahme des Skelettgehalts von Gr2 auf Gr3 mit dem Übergang von der LH zur LB nennenswert (Anhang 2).

Innerhalb der beiden Bodenhorizonte, welche in der LH ausgebildet sind zeigen die oxalatlöslichen Metallverbindungen einen geringen Anstieg von 0,783 auf 0,788 % für pedogenes Eisen, von 0,001 auf 0,006 % für pedogenes Mangan und einen erheblichen von 0,573 auf 1,272 % für pedogenes Aluminium mit der Tiefe. Die dithionitlöslichen Bestandteile steigen von 1,292 auf 1,463 % für das Eisen, von 0,003 auf 0,015 % für Mangan und jene des Aluminiums von 0,766 auf 1,385 % an (Anhang 3).

4.1.4 BC-NSW 13 D

Das Vergleichsprofil BC-NSW 13 D (Zone 32 N: 444380 E; 5387280 N) zu dem zuvor beschriebenem Bodenprofil BC-NSW 13 C befindet sich auf etwa 605 m über NHN außerhalb der kleinen Blockhalde, die sich hangoberhalb des Greßbaches zwischen Hundsbach und Hundseck erstreckt (Anhang 1). Neben Buchen wird die Vegetation durch Tannen und Fichten geprägt; die Krautschicht besteht an diesem Ort aus vereinzelten Vorkommen von Frauenhaarmoos, Sauerklee und Waldschwingel.

Die Bodenform dieses Standortes wurde als gering podsolige Braunerde (Aeh/Bsv/IIilCv) angesprochen, deren LH (Zweiglimmergranit, Sandstein des Buntsandsteins, Löss & LST) über

einer LB (Zersatz des Zweiglimmergranits) entwickelt und von einem mullartigen Moder zur Geländeoberkante hin begrenzt ist. Innerhalb der LH sind ein Aeh-Horizont mit einer Mächtigkeit von 5 cm und ein Bfv-Horizont mit einer solchen von 35 cm wie in der Abbildung 7 ersichtlich ausgebildet. Neben einer leichten Variation der Färbung von hellbraun zu dunkelbraun und einer Abnahme der Durchwurzelung von Wg2/Wf2 auf Wg0/Wf0 bei geringfügiger Erhöhung des Skelettgehaltes von Gr2 auf Gr3 ist ein invariables Kohärentgefüge bei Änderung der Korngrößenzusammensetzung von Sl3 zu Sl4 mit dem Wechsel von LH zur LB vorzufinden (Anhang 2).

Innerhalb der beiden in der LH entwickelten Bodenhorizonte zeigen die oxalatlöslichen Metallverbindungen eine negative Tendenz von 0,279 auf 0,253 % für pedogenes Eisen, einer gleichbleibenden Anteil von 0,001 % für pedogenes Mangan sowie einen geringfügigen Anstieg von 0,137 auf 0,138 % für pedogenes Aluminium mit der Tiefe. Die dithionitlöslichen Bestandteile zeigen einen gleichartigen Trend; dessen Anteil sinkt von 0,611 auf 0,495 % für das Eisen, stagniert bei 0,002 % für Mangan und steigt lediglich für Aluminium von 0,137 auf 0,141 % geringfügig an (Anhang 3).

Abb. 7: Lichtbild des podsoligen Braunerdeprofils BC-NSW 13 D (eigene Aufnahme).

4.1.5 BC-NSW 13 E

Das Profil BC-NSW 13 E (Zone 32N: 450450 E; 5389840 N) befindet sich auf etwa 690 m über
NHN hangoberhalb des Nordufers der Schwarzenbachtalsperre (Anhang 1). Neben Buchen setzt
sich die Bestockung insbesondere aus Fichten zusammen, welche über einer Krautschicht von
Frauenhaarmoos sowie Waldschwingel heranwachsen.

Die Bodenform dieses Standortes wurde als eine Lockerbraunerde (Ah/Bfv/IIilCv)
angesprochen, deren LH (Zweiglimmergranit, Sandstein des Buntsandsteins, Löss & LST) über
einer LB (Zersatz aus Zweiglimmergranit und Sandstein des Buntsandsteins) entwickelt ist und von
einem der Hauptlage aufliegenden, feinhumusarmen Moder begrenzt wird. Wie in der Abbildung 8
zu betrachten, sind innerhalb der LH ein Aeh-Horizont mit einer Mächtigkeit von etwa 15 cm und
ein Bfv-Horizont bis in die absolute Tiefe von 65 cm ausgebildet. Neben einem Farbumschlag von

einem ockerstichigen Braun zu einem
violettstichigen Braungrau zeigt sich bei
einem unverändertem Kohärentgefüge
eine Abnahme der Durchwurzelung von
Wg1/Wf2 zu Wg0/Wf0. Weiterhin ist insbesondere eine markante Erhöhung des
Skelettgehalts von Gr3 auf Gr6 bei ähnlicher Korngrößenverteilung des Feinbodens innerhalb der Kategorie Ls4 im
Übergangsbereich von der LH zur LB von
größerer Relevanz (Anhang 2).

Innerhalb der beiden in der LH entwickelten Bodenhorizonte zeigen die oxalatlöslichen Metallverbindungen einen
Anstieg von 0,534 auf 0,658 % für pedogenes Eisen, von 0,001 auf 0,002 % für
pedogenes Mangan und von 0,172 auf
0,396 % für pedogenes Aluminium mit zu-
nehmender Tiefe. Die dithionitlöslichen
Bestandteile steigen von 0,932 auf 1,151
% für das Eisen, von 0,003 auf 0,006 %
für Mangan und von 0,216 auf 0,466 %
für Aluminium an (Anhang 3).

Abb. 8: Lichtbild des Lockerbraunerdeprofils BC-NSW 13 E (eigene Aufnahme).

4.2 Vergleich der Ergebnisse

Auf Grundlage der erhobenen bodenphysikalischen wie auch bodenchemischen Parameter ergibt sich der Bodentyp einer Lockerbraunerde für die Profile BC-NSW 13 A, BC-NSW 13 C und BC-NSW 13 E wie er gemäß der Ad-hoc Arbeitsgruppe Boden in der Kartieranleitung der 5. Auflage abgegrenzt wurde (AG Boden 2005: 216).

Das Trockenraumgewicht des diagnostischen Horizontes lässt sich entgegen den üblichen Werten von > 1,19 g/cm³ für sandiges und lehmiges Bodensubstrat in einem Wertebereich von < 1 g/cm³ verorten, was der lockeren Lagerung der auflagernden Ah-Horizonte entspricht (Blum 2007: 58; Hintermaier-Erhard & Zech 1997: 37). Eine definitionsgemäße Unterschreitung des durch die AG Boden 2005 festgelegten Grenzwertes von 0,8 g/cm³ für das Trockenraumgewicht des Bfv-Horizontes findet hingegen lediglich in Profil A statt, was infolge der hohen Differenz zu den weiteren Messwerten jedoch eher auf die standörtliche Besonderheit eines zum Erhebungszeitpunkt nicht zur Gänze aufgetauten Bodens und einer damit einhergehenden Volumenzunahme des Wassers um bis zu 10 % infolge kryostatischen Drucks in Zusammenhang mit einen Probenahmefehler zurückgeführt werden könnte (Leser 2009: 202; AG Boden 2005: 216). Das Trockenraumgewicht der Verbraunungshorizonte, welche in den Vergleichsprofilen der Braunerden vorliegen, kann hingegen auf Werte > 1,1 g/cm³ beschränkt werden, womit diese dem unteren Segment des gängigen Wertebereichs für Mineralböden zugeordnet werden können (Anhang 3).

Entgegen diesem Charakteristikum kann ein definitionsgemäßes Gesamtporenvolumen von über 60 % bei allen locker gelagerten Hauptlagen von Profilen mit tixotrophen Eigenschaften festgestellt werden; jenes der Vergleichsprofile beläuft sich für den Verbraunungshorizont – wie in Abbildung 9 ersichlich – lediglich auf Werte, welche einen Anteil von 50 % nur geringfügig überschreiten (Anhang 2).

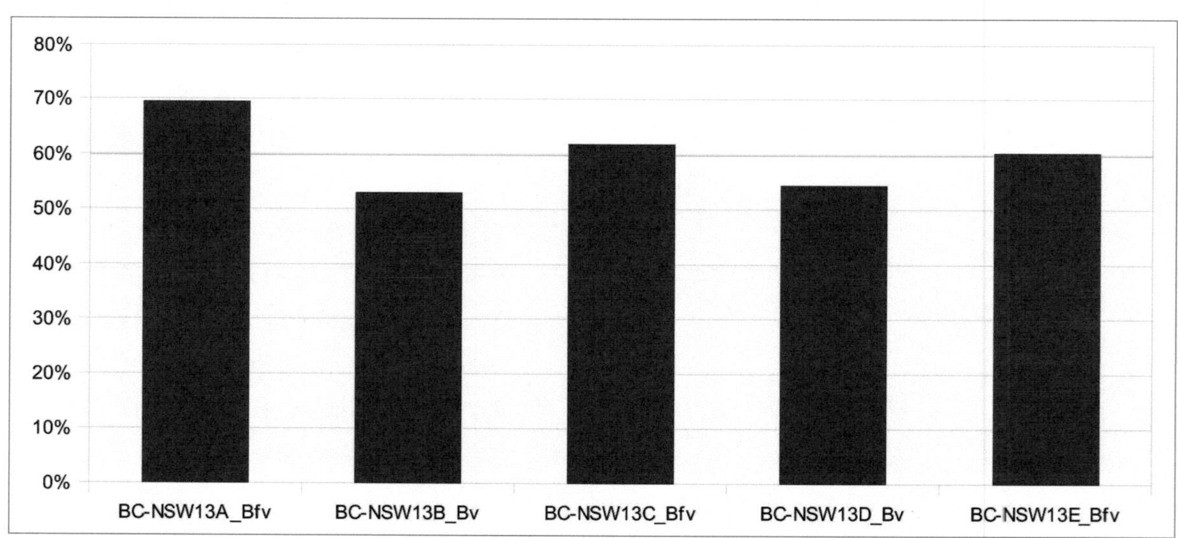

Abb. 9: Porenvolumina der untersuchten Verbraunungshorizonte (eigene Darstellung).

Die ermittelten pH-Werte der beprobten Bodenhorizonte lassen hingegen keine derartige Differenzierung zwischen den Lockerbraunerden und ihren Vergleichsprofilen zu. Obgleich eine Differenz von etwa 0,5 beim Vergleich der diagnostischen Horizonte von BC-NSW 13 D und BC-NSW 13 E festgestellt werden kann, die bereits durch VÖLKEL als mögliches Unterscheidungskriterium diskutiert wurde, zeigt sich im Vergleich von BC-NSW 13 A und BC-NSW 13 B kein derartiger Unterschied; der pH-Wert der Lockerbraunerde ist in diesem Falle sogar geringfügig niedriger als jener des Bodenhorizontes innerhalb des Vergleichsprofils (Anhang 3).

Die Unterschreitung eines pH-Wertes von etwa 3,8 in den Ah-Horizonten und von 4,5 in den Verbraunungshorizonten ist indes als eine generelle Folge von basenarmen Ausgangssubstrat in Kombination mit den niederschlagsreichen Klimaverhältnissen unabhängig vom ausgebildeten Bodentyp zu betrachten. Erwähnenswert hierbei ist jedoch der Unterschied in der optischen Erscheinung; während bei den Vergleichsprofilen Podsolierungserscheinungen im humosen Oberbodenhorizont ersichtlich sind, welche durch Eluviation im Zuge der niedrigen pH-Werte eintreten, zeigen die Lockerbraunerdeprofile, wie bereits durch SCHÖNHALS beschrieben, zumindest keine optischen Anzeichen eines solchen Prozesses, sodass bei diesen aufgrund des spezifischen Verlaufs der pH-Werte von einer sogenannten Kryptopodsolierung ausgegangen werden kann (MEYER & SAKR 1970a: 52; WIECHMANN 2000: 6). Die Verbraunungshorizonte sind bei den Vergleichsprofilen desweiteren eher in gelbbraunen Farbtönen ausgeprägt, wohingegen die diagnostischen Bodenhorizonte der Lockerbraunerden in einem leutenden ockerbraun in Erscheinung treten.

Eine Tendenz lässt sich dem entgegen in den Anteilen der organischen Substanz erkennen, die mit Werten zwischen 3,0 und etwa 5,5 % in den unterschiedlichen Verbraunungshorizonten einen generell hohen Prozentsatz einnimmt, der vermutlich erneut auf Verlagerungsvorgänge der Podsolierung zurückzuführen sein dürfte (WIECHMANN 2000: 3). Die Werte der podsoligen Braunerden lassen sich dabei an das untere Ende, jene der Lockerbraunerden am oberen Ende dieser Wertespanne ansiedeln, wobei eine Grenze bei einem Anteil der organischen Substanz von 4 % gezogen werden kann. Eine Korrelation mit der Aktivität der bodenbewohnenden Mikroorganismen lässt sich nicht erkennen, da entgegen einer üblichen Abnahme der Aktivität mit der Abnahme des pH-Wertes im Fall der untersuchten Bodenprofile selbst bei höheren Werten der Anteil der organischen Substanz das obere Ende der Spanne erreicht (SCHEFFER & SCHACHTSCHABEL 2010: 104).

Bei der Analyse der Korngrößenverteilung lässt sich erneut keine einheitliche Differenzierung zwischen den verschiedenen Bodentypen feststellen. Generell sind im Untersuchungsgebiet, wie in Abbildung 10 auf der folgenden Seite visualisiert, Böden aus sandigem Lehm bzw. lehmigen Sand verbreitet, bei denen mitunter eine leichte Lessivierung im Zuge der Korngrößenanalyse ersichtlich wird. Dieser optisch im Gelände nicht wahrnehmbare Prozess steht jedoch eher im Zusammenhang mit der generellen Verlagerungstendenz, welche auf die hohen Niederschlagsmengen zurückzuführen ist, und beschreibt auch aufgrund seiner vergleichsweise geringen Ausprägung entgegen der Podsolierung nicht den charakteristischen Prozess der Pedogenese (Anhang 3).

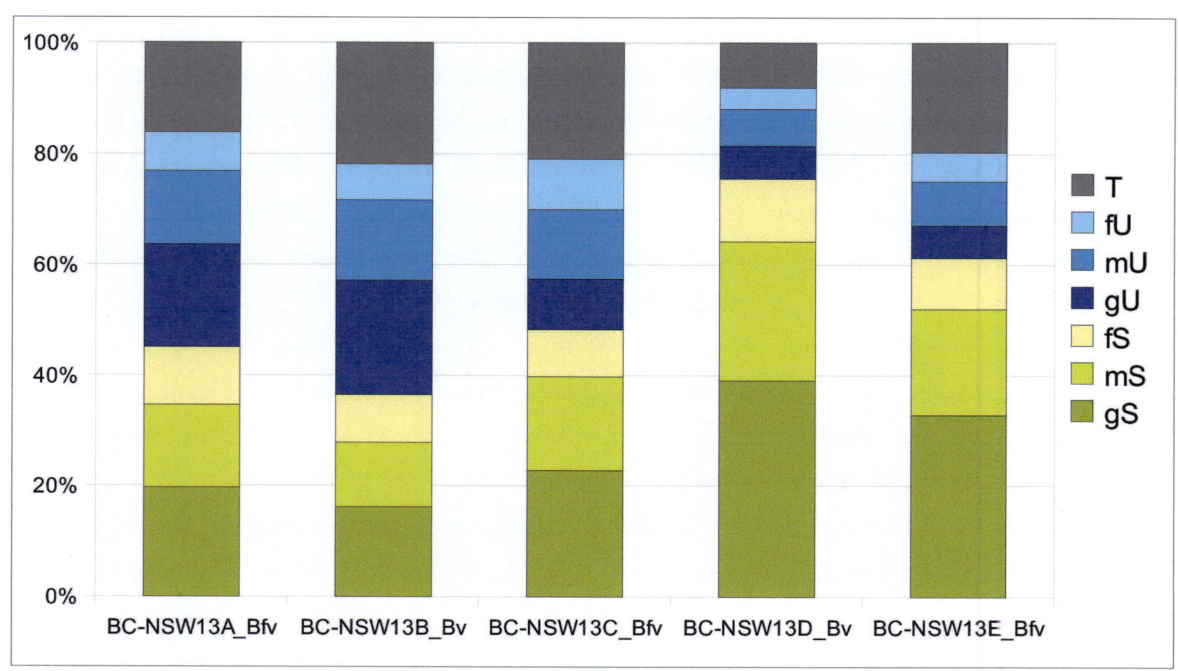

Abb. 10: Korngrößenzusammensetzung der untersuchten Verbraunungshorizonte (eigene Darstellung).

Jedoch bildet sich ein Unterschied unabhängig vom entwickelten Bodentyp in der Korngrößenverteilung in Hinblick auf die Lage der einzelnen Profile heraus. Während sich bei den Proben aus Bodenhorizonten der Profilnummern BC-NSW 13 A und BC-NSW 13 B eine ausgeprägte Schluffkomponente in Größenordnungen von über einem Drittel an der Gesamtheit mit einem relativen Maximum im Bereich des Grobschluff feststellen lässt, geht diese innerhalb der restlichen beprobten Bodenhorizonte großteils deutlich auf einen relativen Anteil von etwa einem Fünftel mit einem Maximum im Mittelschluff zurück. Entsprechend tritt in den Bodenprofilen der Kennung BC-NSW 13 C bis BC-NSW 13 E der Anteil der Sandfraktion in der Hauptlage deutlich hervor, was durch die solifluidale Aufarbeitung von Buntsandsteinzersatz innerhalb dieser näher am Deckgebirge gelegenen Profile erklärt werden kann, da sich oftmals selbst größere Exemplare eingeregelten Buntsandsteinzersatzes in der Hauptlage vorfinden lassen.

Eindeutige Tendenzen lassen sich dem entgegen wiederum aus der Verteilung und der Quantifizierung bestimmter pedogener Metalloxide erkennen, wie sie in Abbildung 11 auf Seite 26 verzeichnet sind (Anhang 3). Während sich die Fe_o-Gehalte der Vergleichsprofile lediglich in einem Bereich von 0,25 bis unter 0,5 % bewegen, sind die Gehalte innerhalb der als Lockerbraunerde angesprochenen Profile mit 0,65 bis 0,8 % deutlich ausgeprägter. Ebenso lässt sich eine Regelhaftigkeit im Verlauf innerhalb der Profile erkennen. Während innerhalb der Lockerbraunerdeprofile stets eine minimale bis größtenteils starke Zunahme des Fe_o-Gehaltes mit dem Übergang vom humosen Oberboden zum locker gelagerten Verbraunungshorizont zu verzeichnen ist, ergibt sich bei den podsoligen Braunerden durch eine Abnahme des relativen Anteils mit der Tiefe innerhalb der Hauptlage ein konträres Bild. Zudem zeigt sich bei dem Konzentrationsverlauf innerhalb des Profils BC-NSW 13 E ein sprunghafter Rückgang des Gehaltes an oxalatlösichem Eisen mit dem Übergang von der Haupt- zur Basislage, wodurch sich insbesondere die Bedeutung von Lagewechseln in Bezug auf eine pedogenetische Interpretationen zeigt (vgl. VÖLKEL 1995: 118).

Bei den Gehalten des oxalatlöslichen Mangans ist dem entgegen keine Korrellation mit dem entsprechenden Bodentypen erkennbar, einzig eine generelle Zunahme mit dem Übergang vom humosen Oberboden in verbraunte Bereiche scheint regelhaft. Selbst die erhöhten Gehalte innerhalb des Profils BC-NSW 13 A sowie vor allem im zugehörigen Vergleichsprofil BC-NSW 13 B liefern keinen interpretatorischen Ansatz zu einer Differenzierung zwischen den Lockerbraunerden und den umgebenden Bodentypen aufgrund konstant niedriger Gehalte von bis zu einem halben Gewichtsprozent in den Profilen BC-NSW 13 C bis BC-NSW 13 E, sondern lassen einzig erhöhte Mangangehalte in den Böden gewässerparallel zum Gertelsbach erkennen.

Auch die Gehalte des oxalatlöslichen Aluminiums lassen keine bodentypspezifischen Größenordnungen voneinander abgrenzen. Vielmehr scheint eine generelle Zunahme des relativen Anteils von humosen Oberbodenhorizonten zu den tiefergelegenen Verbraunungshorizonten gegeben zu sein, der bei den Lockerbraunerden einen signifikanten Anstieg um 100 % ausmacht, während sich ein solcher bei den podsoligen Braunerden auf einen wesentlich geringeren Prozentsatz beläuft bzw. nahezu unveränderte Werte im Tiefenverlauf zeigt.

Bei der Verteilung der dithionitlöslichen Eisenverbindungen zeigt sich eine ähnliche Tendenz wie bei dem oxalatlöslichen Anteil dieser Eisenverbindungen. Während die Werte der Vergleichsprofile mit der Tiefe rückläufig bis nahezu stagnierend sind, ergibt sich für den Bodentyp der Lockerbraunerde im Untersuchungsgebiet erneut eine Zunahme des relativen Anteils bei einem Wechsel vom Ah- zum Bfv-Horizont. Dabei ließe sich eine Grenze bei einem Wert von etwa einem Prozent zur Abgrenzung definieren, der von den thixotrophen Bodenhorizonten der Lockerbraunerden überschritten, von den restlichen Verbraunungshorizonten jedoch unterschritten wird.

Ebenso zeigt auch der dithionitlösliche Anteil des Mangans eine ähnliche horizont- und profilbezogene Verteilung wie der oxalatlösliche Anteil dieser Metallverbindung. Auch hier weisen erneut die Profile am Gertelsbach die höchsten Gehalte auf.

Wiederum ergibt sich bei den relativen Gehalten des dithionitlöslichen Anteils der Aluminiumverbindungen keine eindeutige Tendenz zur Abgrenzung der Lockerbraunerde; während innerhalb des Verbraunungshorizontes des Profils BC-NSW 13 C zwar ein Wert von fast 1,4 % erreicht wird, wohingegen der des Vergleichsprofils BC-NSW 13 D sich lediglich auf etwa ein Zehntel davon beläuft, bewegen sich die relativen Anteile der Verbraunungshorizonte für alle weiteren Profile in einer Größenordnung ohne ausweisbare Abgrenzung, da der Wertebereich von 0,4 bis 0,7 % allen Bodentypen ohne absehbaren Trend gemein ist.

Relativierend in die Gesamtbetrachtung sollten die erhöhten Mangangehalte insbesondere des Profils BC-NSW 13 B sowie die erhöhten Gehalte der Aluminiumverbindungen im Profil BC-NSW 13 C einfließen, da diese hohe Abweichungen zu den anderen bodenchemischen Ergebnissen aufweisen und vermutlich eher lokale Eigenarten als ein bodentypspezifisches Kriterium darstellen.

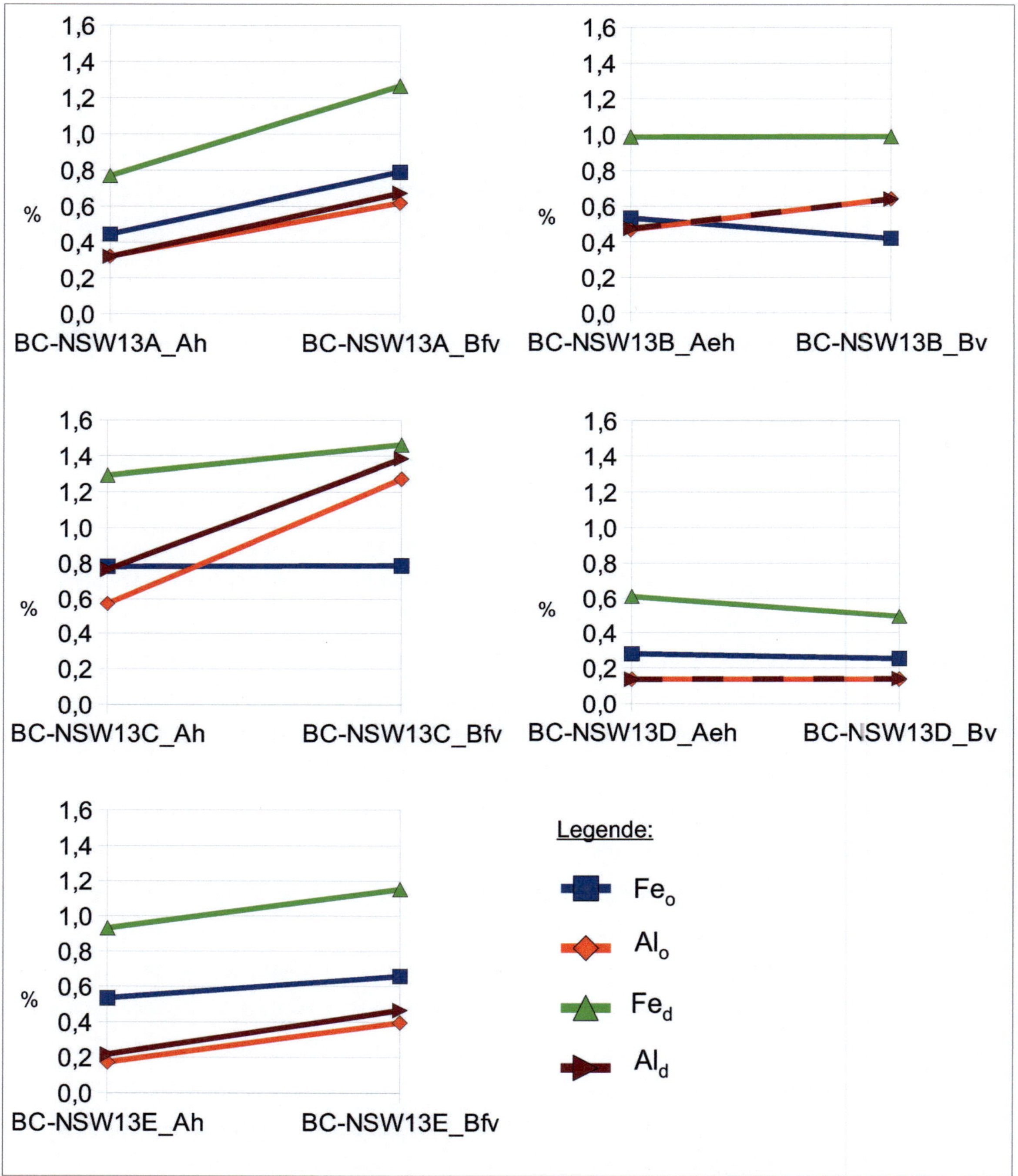

Abb. 11: Verlauf der pedogenen Metalloxide innerhalb der LH der aufgenommenen Bodenprofile BC-NSW 13 A bis BC-NSW 13 E (eigene Darstellung).

5 Interpretation und Diskussion

Das die untersuchten Bodentypen, welche eine überaus locker gelagerte Hauptlage mit hoher Tixotrophie aufweisen, sich in mächtigen äolischen Ablagerungen von lokal verwehtem Basaltdetritus oder einer Tephra entwickelt haben, kann zunächst ausgeschlossen werden.

Eine Genese der Lockerbraunerden aus olivinhaltigem Basaltzersatz, wie sie in Publikationen anderer Autoren etwa innerhalb des Vogelsbergs nachgewiesen wurde, kann für das Untersuchungsgebiet im Nordschwarzwald zunächst ausgeschlossen werden, da über den gesamten Verlauf des Höhenzuges keine Basaltmassen anstehen (vgl. METZ 1977: 12). Einzig der südsüdwestlich in etwa 80 km Entfernung befindliche und aus basaltischen Intrusivmassen hervorgegangene Kaiserstuhl würde als Fördergebiet von äolisch verlagertem Basaltdetritus in Frage kommen, jedoch ist ein solcher Prozess aufgrund der relativ großen Entfernung, der vorherrschenden Hauptwindrichtung gen Osten und der Lokalisierung der Lockerbraunerden in exponierten Lagen als unwahrscheinlich zu bewerten (vgl. SAKR & MEYER 1970: 4; METZ 1977: 12).

Auch eine Entwicklung innerhalb mächtiger, äolisch akkumulierter Tephra kann im Untersuchungsgebiet nicht nachvollzogen werden, da die einzigen großflächigen Ablagerungen derartiger Vulkanaschen, die im Raum des Schwarzwaldes abgelagert wurden und in die quartäre Bodenbildung eingehen konnten, dem Laacher See Vulkan entstammen, welcher gegen Ende des Alleröd vor etwa 12.900 Jahren in der östlichen Eifel ausbrach (MEYER & SAKR 1970a: 51; SCHMINCKE 2000: 173). Die vulkanischen Gläser beschränken sich in Sedimenten innerhalb der Hochlagen des Nordschwarzwaldes jedoch auf eine Mächtigkeit von etwa 2 cm, sodass eine gänzliche solifluidale Aufarbeitung dieser in den Hangbereichen mit weiterem Substrat während des Periglazials zu einer Hauptlage mit Sicherheit stattgefunden hat (SCHAMMEL 1991: 116; LOTTER & BIRKS 1993: 266). Obgleich die vorgefundenen Bodenprofile nur in kleinen Arealen zumeist nahe Blockhalden vorzufinden sind, kann auch die Akkumulation größerer Mächtigkeiten infolge von Erosion der vulkanischen Asche und folgender Ausbildung einer Lockerbraunerde aus Laacher-See-Tephra ausgeschlossen werden, da sich die fraglichen Profile eben nicht in kleinräumigen Mulden oder Senkenbereichen sondern an relativ exponierten Standorten entwickelt haben (STÖHR 1963: 332).

Desweiteren konnte auch über eine Schwermineralanalyse der lediglich geringe Anteil an Laacher-See-Tephra in Bereichen dieser tixotrophen Böden nachgewiesen werden (*freundl. mündl. Mitt. Prof. Dr. Thiemeyer*). Bei einem Großteil der Standorte konnten die entsprechenden vulkanischen Schwerminerale braune Hornblende, Titanit sowie das Pyroxen Augit dabei überhaupt nicht in vollem Spektrum in der Feinsand-Fraktion der Hauptlage vorgefunden werden (WINDHEUSER 1977: 18f.). Auch wenn dieser Negativnachweis nach SAKR & MEYER nicht bedeutet, dass keine entsprechenden Mächtigkeiten von Vulkanascheablagerungen im Untersuchungsgebiet zur Entstehungszeit vorzufinden waren, so kann aufgrund der Position von den hohlraumreichen Bodentypen in exponierter Lage und infolge der von SCHMINCKE beschriebenen, flächenhaft geringmächtigen Ascheablagerung im Bereich des Schwarzwaldes ebenso ein lediglich geringfügiger

Anteil der Allophane vermutet werden, welche der Verwitterung von Tuffen entstammen würden, sodass sich eine Quantifizierung im Zuge der vorliegenden Arbeit erübrigt (SAKR & MEYER 1970: 26).

Auch die Betrachtung der bodenphysikalischen Parameter liefert außer des zutreffenden Kriteriums einer definitionsgemäßen Unterscheidung der einzelnen Braunerdevarietäten über das Porenvolumen keinen weiteren Ansatz. Eine Erklärung für die unabhängig vom ausgebildeten Bodentyp auftretenden erhöhten Schluffgehalte innerhalb der in Nähe zum Westabfall des Schwarzwaldes gelegenen Bodenprofile findet sich indes infolge der geringmächtigen Akkumulation von Löss aus der Oberrheinebene. Denn ebenso wie die Schluffkomponente der in ostexponierter Hanglage befindlichen Profile BC-NSW 13 A und BC-NSW 13 B weist dieses Lockergestein ein Korngrößenmaximum bei Grobschluff auf und wurde nachweislich außer in größeren Mächtigkeiten im Bereich der Schwarzwaldrandverwerfung auch zu einem wesentlich geringeren Teil in den Hochlagen des Gebirges akkumuliert (METZ 1977: 59; JAHN ET AL. 1994: 28). Die restlichen Bodenprofile BC-NSW 13 C bis BC-NSW 13 E, welche bei ähnlicher hypsometrischer Verteilung eine größere Entfernung zum Oberrheingraben und keine Lage in einer Leeposition aufweisen, beinhalten eine geringere Schluffkomponente geringerer Korngröße welche auch aus lokalen Verlagerungen entstammen kann, da solifluidal aufgearbeitete Verwitterungsreste des Ausgangsgesteins innerhalb dieser Korngrößenspektren stärker in Erscheinung treten (JAHN ET AL. 1994: 28).

Auch ein Vergleich der Ergebnisse aus der Korngrößenanalyse mit denen der bodenchemischen Parameter lässt keinen Ansatz einer pedogenetischen Erklärung der Existenz von Lockerbraunerden im Nordschwarzwald zu. Ein Zusammenhang zwischen dem relativen Tonanteil in der Hauptlage und jenem der pedogenen Oxide ist nicht festzustellen, woraus auf einen lediglich geringen Anteil zumindest an sekundär chloritisierten Dreischicht-Tonmineralen geschlossen werden kann, da diese Komplexbildner für Eisen-, Aluminium und Manganoxide darstellen und entsprechend ein hohlraumreiches Gefüge erzeugen könnten (VÖLKEL 1995: 121).

Deutlich wird jedoch ein Zusammenhang bei der Betrachtung der differenzierungsrelevanten bodenchemischen Charakteristika untereinander. Zwischen dem Anteil der organischen Substanz und der relativen Häufigkeit der pedogenen Metalloxide innerhalb der beiden betrachteten Braunerdevarietäten ergeben sich Parallelen, auch wenn sich bei diesen kein direkt proportionaler Zusammenhang erkennen lässt. Augenscheinlich gehen mit den erhöhten Gehalten der organischen Substanz innerhalb des Verbraunungshorizontes der Lockerbraunerden erhöhte Gehalte der pedogenen Metalloxide im Allgemeinen wie dem des schlecht kristallisierten Eisens im Speziellen infolge der Komplexbildung durch unterschiedliche spezifische Ladung einher (vgl. MEYER & SAKR 1970a: 55ff.; VÖLKEL 1995: 122; MAHR 1998: 271). Die entstandenen Metallhumuskomplexe, deren organische Substanz zu einem Teil aus langkettigen Polysacchariden aufgebaut ist, können durch ihre Verkettung untereinander und folgender Generierung einer größeren Oberfläche ebenso ein Gefüge mit entsprechend größerem Porenvolumen vergleichbar mit jenem eines humosen Oberbodenhorizontes erzeugen (vgl. VÖLKEL 1995: 122f.).

Daneben ist auch die Ausbildung eines entsprechend hohen Porenvolumens der Lockerbraunerden durch kryogene Volumenänderung nicht auszuschließen. Entsprechend der geringen Evapotranspiration sowie der relativ hohen Luftfeuchtigkeit, welche sich durch die geringen Temperaturen zu einem Großteil des Jahres in den Hochlagen des Nordschwarzwaldes ergeben, wird der Boden stetig intensiv durch die hoher Niederschlagsmengen durchfeuchtet (vgl. Brahmer & Feger 1987: 290; Wilmanns 2001: 18). Durch den häufigen Wechsel von Auftau- und Gefrierprozessen, die sich über einen langen Zeitraum hinweg zum Jahresende sowie zu Beginn eines jeden Jahres erneut infolge Temperaturschwankungen nahe dem Gefrierpunkt ergeben, kann es aufgrund des hohen Anteils am Porenraum, welcher mit Bodenwasser gefüllt ist, zu einer ausgeprägten Kammeisbildung kommen (vgl. Brunnacker 1965: 74; Wilmanns 2001: 22). Derartige Erscheinungen sind in geringerer Ausprägung jedoch regelhaft in der gemäßigten Klimazone anzutreffen; nicht einzig die Ausbildung eines hohlraumreichen Gefüges, sondern auch die Stabilisierung dessen über die wärmere Jahreszeit hinweg muss deshalb gegeben sein.

Eine derart gefügestabilisierende Wirkung weisen sowohl die organische Substanz als auch pedogene Metalloxide, welche in Bereichen der Lockerbraunerden in hohen relativen Anteilen auftreten, ungeachtet ihrer möglichen Gefügeausbildung per se auf (vgl. Völkel 1995: 122f.; Mahr 1998: 270). Diese Wirkung der Metallverbindungen ist insbesondere im Zusammenhang mit der Illuviation und der damit einhergehenden Ausbildung eines Kittgefüges im Rahmen von Podsolierungsprozessen bekannt (Scheffer & Schachtschabel 2010: 289). Neben den oxalatlöslichen Aluminiumverbindungen, auf Grundlage deren relativer Anteile keine Unterscheidung der beiden Braunerdevarietäten möglich war, können insbesondere schlecht kristallisierte Eisenoxide in Form der leuchtend orangefarbenen Ferrihydritminerale, die bei den Lockerbraunerden in Relation zu den podsoligen Braunerden ungleich höhere Werte aufweisen, bereits mit einem relativen Anteil von wenigen Prozent die Oberflächen des Bodensubstrates vollständig umhüllen und auf diese Weise die physikalischen wie auch chemischen Eigenschaften des Bodens in einem überproportionalem Maße beeinflussen (vgl. Childs 1992: 444f.; Mahr 1998: 271). Ein Vergleich der pedogenen Gesamtgehalte der Eisenoxide verdeutlicht zudem, dass es sich nicht um interfraktionelle Verschiebungen handeln kann; so stehen den Gesamtgehalten an Eisenoxiden von bis zu 1,4 % bei den podsoligen Braunerden erheblich höhere Anteile von teilweise über 2,2 % bei den Lockerbraunerden innerhalb des Verbraunungshorizontes gegenüber.

Ein Problem stellt jedoch die Rekonstruktion des gesamten Faktorenkomplexes der pedogenetischen Prozesse dar, welche einen derartigen Bodentypen mit den erwähnten pedochemischen Charakteristika entstehen ließen (vgl. Völkel 1995: 122; Mahr 1998: 271). Die Differenz zwischen den Eisenoxidgehalten innerhalb der unterschiedlichen Braunerdevarietäten nicht einzig im Verbraunungshorizont sondern in der gesamten Hauptlage legt nahe, dass bei den Lockerbraunerden eine Zufuhr der Metalloxide von außerhalb stattgefunden haben muss. Gegen einen äolischen Eintrag eisenhaltiger Stäube im Sinne von Lokalverwehungen, wie ihn Völkel und Mahr für die Lockerbraunerden des Bayerischen Waldes diskutieren, spricht das geringe Verbreitungs-areal dieser Böden im Untersuchungsgebiet des Nordschwarzwaldes auf wenigen Quadratmetern sowie ein teilweises Vorkommen an exponierten Hanglagen, an welchen sich für gewöhlich keine Deflationsgebiete äolischer Sedimente ausbilden.

Diese Kriterien der Profillage im Gelände und der Kleinräumigkeit lassen ebenso eine Auf-arbeitung mesozoisch-tertiären Verwitterungsmaterials in großen Mengen innerhalb der perigla-ziären Lagen, wie bereits durch BRUNNACKER für die Lockerbraunerden des Bayerischen Waldes dis-kutiert, als unwahrscheinlich erscheinen, obgleich durch die Einarbeitung von derartigem Substrat, welches eine intensive chemische Vorverwitterung erfuhr, die Färbung des Verwitterungshorizon-tes wie auch der hohe Anteil an pedogenen Oxiden eine Erklärung finden würde (vgl. BRUNNACKER 1965: 68). Doch ist ein derartiges Ausgangssubstrat der Bodenbildung ohnehin durch eine "konsequente tektonische Hebung und dazu parallel verlaufende Erosion während des Tertiärs und des Quartärs bis auf kaum zu diagnostizierende Reste abgetragen" (JAHN ET AL. 1994: 31).

Höhere Gehalte im Verbraunungshorizont in Relation zum humosen Oberboden innerhalb der Lockerbraunerden sprechen dem entgegen eher für den Prozess der Podsolierung, deren optisch herausragendes Merkmal eines Eluvialhorizontes durch den hohen Anteil an Organika innerhalb des Oberbodens verschleiert wird (vgl. MAHR 1998: 272; SCHEFFER & SCHACHTSCHABEL 2010: 289). Aufgrund der bereits erwähnten hohen Niederschlagsmengen und geringen Temperaturen im Zusammenhang mit einer geringen Evapotranspiration ergibt sich im Untersuchungsgebiet jedoch zudem potentiell eine Bodenwassersättigung mit möglichem Lateralfluss bei weiterer Wasserzu-fuhr in der Folge (vgl. BRAHMER & FEGER 1987: 290). Zwar würden gegen einen solchen hangparalle-len Abfluss die grusigen Basislagen mit der ihr eigenen, hohen hydraulischen Leitfähigkeit sprechen. Doch lässt das Vorkommen der Lockerbraunerden am Rande von Blockhalden die Ver-mutung zu, dass der Abfluss sich in den feinsubstratreichen Bereichen zwischen diesem blockigen, an der Geländeoberkante befindlichen Zersatz sammelt und es entsprechend zu einer infil-trierenden Wassermenge über der hydraulischen Leitfähigkeit der Basislage und in der Folge zu einem Lateralfluss kommt (vgl. PRIEHÄUSSER 1966: 186ff.; BRAHMER & FEGER 1987: 290ff.; MAHR 1998: 274f.). Ein weiteres Argument für eine solche laterale Podsolierung ergibt sich neben dieser Differ-enz der Gesamtgehalte aus der Stagnation bis hin zu einem Rückgang bei den Anteilen der pedogenen Oxide mit der Tiefe innerhalb der podsoligen Braunerden, welche sich in der Umge-bung der Lockerbraunerden befinden. Ein Zwischenabfluss ausgehend von den podsoligen Braun-erden würde mit einer Abreicherung an pedogenen Metalloxiden aus dem wassergesättigten Bereich einhergehen, die folgend im Bereich der Lockerbraunerden wieder innerhalb des Verbrau-nungshorizontes ausfällen. Durch eine erneute Abnahme der Wassersättigung am Rande der Blockfelder aufgrund des höheren Anteils an Feinsubstrat pro Flächeneinheit ergibt sich ein sauer-stoffreiches Milieu, welches in der Folge die Oxidation der Metalle begünstigen würde.

Neben diesen gegenwärtigen Verlagerungsprozessen wäre weiterhin auch ein Lateralfluss bereits zur Zeit des Periglazials möglich gewesen. Die Permafrostschicht unterhalb der Auftauzone stellte eine undurchlässige Barriere dar, sodass infiltriertes Wasser sich im darüberliegenden Bereich sammelte und folglich den "active layer" ausbildete, aus dem die heutige Hauptlage entstand. Innerhalb dieser sich langsam hangabwärts bewegenden, wassergesättigten Lage wäre eine laterale Verlagerung der pedogenen Metalle denkbar (vgl. VÖLKEL 1995: 124).

Ungeachtet des Zeitpunktes scheint die Existenz der Lockerbraunerden im Nordschwarzwald schlussendlich auf derartige Anreicherungen an Sesquioxiden wie auch organische Substanz durch vertikale wie zeitweise auch laterale Podsolierungsprozesse im weiteren Sinne einherzugehen. Zumindest deuten die erhobenen Gesamtgehalte der bodenchemischen Parameter für das Untersuchungsgebiet im Nordschwarzwald auf derartige Prozesse innerhalb eines möglicherweise weitaus mehr klimatische Kriteren umfassenden Faktorenkomplexes hin (vgl. MAHR 1998: 278).

Eine Interpretation der entsprechenden Profile als Ockererden, welche für den Naturraum des Schwarzwaldes unter anderem von FIEDLER und JAHN beschrieben wurden, kann trotz ähnlicher bodenphysikalischer wie auch -chemischer Kriterien, welche allerdings zum Großteil der Definition der Lockerbraunerde entlehnt sind, ebenso nicht nachvollzogen werden (vgl. FIEDLER & JAHN 2005: 5ff.). Obgleich offenbar auch laterale Fließprozesse für die Genese der Ockererden verantwortlich sind, ist deren Profilaufbau sowie folgend auch Pedogenese nicht mit den in dieser Arbeit behandelten Bodenprofilen mit locker gelagertem Verbraunungshorizont in Einklang zu bringen. Neben einer nicht vorzufindenden impermeablen Schicht, welche die lateralen Fließprozesse begünstigt, ist auch keine Zunahme der organischen Substanz mit der Tiefe zu verzeichnen (vgl. FIEDLER & JAHN 2005: 2f.). Schließlich entspricht die skizzierte Bodengesellschaft mit hangoberhalb gelegenen Stauwasserböden in Ausbildung eines Stagnogleys nicht den Verhältnissen der innerhalb dieser Arbeit näher betrachteten und diskutierten Standorte (vgl. FIEDLER & JAHN 2005: 3).

6 Zusammenfassung und Ausblick

Im Rahmen dieser Arbeit wurden Lockerbraunerden sowie anteilig Vergleichsprofile innerhalb von drei im Grindenschwarzwald gelegenen Untersuchungsgebieten auf ihre Verbreitung sowie ihren bodenphysikalischen wie auch -chemischen Charakteristika analysiert und ein Versuch der Identifizierung jener pedogenetischen Prozesse unternommen, welche prägend für die Lockerbraunerden des Nordschwarzwaldes sind.

Infolge eines Vergleichs der Lockerbraunerdeprofile mit den Vergleichsprofilen der podsoligen Braunerden ergeben sich für Erstere neben des erhöhten Porenraumvolumens von > 60 % signifikante Unterscheidungsmerkmale durch einen tiefreichenden, hohen Anteil an organischer Substanz sowie aller Ansicht nach damit einhergehend ein hoher Anteil an oxalatlöslichen Eisen im Verbraunungshorizont. Diese schlecht kristallisierten Eisenoxide in Form von Ferrihycritmineralen bilden die Grundlage für die intensive Ockerfärbung des locker gelagerten Oberbodens sowie für die Stabilisierung des Aggregatgefüges, das unter anderem auf klimatische Prozesse zurückführbar scheint.

Insgesamt zeigen die pedogenen Metalloxide innerhalb des Substrats der Lockerbraunerden zudem eine Zunahme mit der Tiefe, während sie bei den podsoligen Braunerden stagnierend bis rückläufig sind. Diese Anreicherung im Verbraunungshorizont scheint entweder auf äolischem Eintrag oder einer mehrdimensionalen Podsolierung, welche sich aus vertikalen und lateralen Fließprozessen ergibt, zu beruhen. Entsprechende Bodenwasserverhältnisse sind im Untersuchungsgebiet nördlich der Hornisgrinde – abseits eines möglichen Auftretens dieser bereits zur Zeit des Periglazial – zumindest rezent stellenweise und periodisch gegeben. Eine Bilanzierung des Bodenwasserhaushalts an den Lockerbraunerdestandorten ist entsprechend denkbar, um die Aussagen der laterale Fließprozesse zu fundieren und die Qualitität dieser Fließprozesse wie auch Quantität über einen Jahresgang zu evaluieren.

Leider lässt sich der vielschichtige Faktorenkomplex, welcher für die Genese der Lockerbraunerden verantwortlich ist, sich nicht bis in das letzte Detail ergründen; so ist nicht auszuschließen das noch weitere, bisher nicht in Betracht gezogene Prozesse die Ausbildung von Lockerbraunerden bedingen. Zumindest ließe sich durch kleinräumige Unterschiede der naturräumlichen Gegebenheiten die geringe Dimension des Auftretens der Lockerbraunerden im Nordschwarzwald erklären.

In der Folge scheint eine Kartierung der kleindimensionalen Verbreitungsgebiete der Lockerbraunerde im Nordschwarzwald angebracht, um den Faktorenkomplex ihrer Entstehung weiter aufzuschlüsseln und eventuell weiterführende pedogenetische Erkenntnisse für die gesamte Bodengesellschaft der Region zu erlangen. Besonderes Augenmerk sollte dabei das Gebiet der Omerskopf-Gneisscholle erfahren, da an einen solchen geologischen Aufbau auch die vermehrte Verbreitung der nicht-allophatischen Lockerbraunerden im Bayerischen Wald gebunden scheint.

7 Quellenverzeichnis

7.1 Literatur

Abo-Rady, M.D.K. (1985): Schwermetalle in Lockerbraunerden in Vogelsberg und Taunus. *Geologisches Jahrbuch Hessen* 113: 229 – 250. Wiesbaden (Selbstverlag).

AG Boden [= AD-HOC ARBEITSGRUPPE BODEN] (2005): Bodenkundliche Kartieranleitung - 5. Auflage. Stuttgart (Schweizerbart).

Akinci, M.C. (1973): Untersuchungen an der organischen Substanz der Lockerbraunerden und der Sauren Braunerden im Hohen Vogelsberg. Gießen (Selbstverlag).

Bargon, E. (1960): Über die Entwicklung von Lockerbraunerden aus Solifluktionsmaterial im Vorderen Odenwald. *Zeitschrift für Pflanzenernährung, Düngung und Bodenkunde* 90 (3): 229 - 243.

Blum, W.E.H. (2007): Bodenkunde in Stichworten – 6. völlig neu bearb. Auflage. Berlin u.a. (Borntraeger).

Brahmer, G. & K.-H. Feger (1987): Auswirkungen unterschiedlicher Fließwege und biogeochemischer Prozesse auf die chemische Zusammensetzung der Hydrosphäre im Schwarzwald. *Mitteilungen der Deutschen Bodenkundlichen Gesellschaft* 55 (I): 289 – 294.

Brunnacker, K. (1965): Die Lockerbraunerde im Bayerischen Wald. *Geologische Blätter von Nordost-Bayern und angrenzende Randgebiete* 15: 65 – 76.

Childs, C.W. (1992): Ferrihydrite. A review of structure, properties, occurrence in relation to soils. *Zeitschrift für Pflanzenernährung und Bodenkunde* 155: 441 – 448.

Fiedler, S. & R. Jahn (2005): Accumulation soils like "Ockererde" – forgotten soil units in soil-classification systems. *Journal of Plant Nutrition and Soil Science* 168: 1 – 8.

Gassmann, G., Hauptmann, A., Hübner, C., Ruthardt, T. & Ü. Yalçin (2005): Forschungen zur keltischen Eisenerzverhüttung in Südwestdeutschland. Stuttgart (Konrad Theiss).

Günther, D. (2010): Der Schwarzwald und seine Umgebung. Sammlung geologischer Führer 102. Stuttgart (Borntraeger).

Hasel, K. (1944): Herrenwies und Hundsbach – Ein Beitrag zur forstlichen Erschließung des nördlichen Schwarzwaldes. Forschungen zur Deutschen Landeskunde 45. Leipzig (Hirzel).

Heinrichs, H., König, N. & R. Schultz (1985): Atom-Absorptions- und Emissionsspektroskopische Bestimmungsmethoden für Haupt- & Spurenelemente in Probelösungen aus Waldökosystemuntersuchungen. Berichte des Forschungszentrums Waldökosysteme/Waldsterben 3.

Hintermaier-Erhard, G. & W. Zech (1997): Wörterbuch der Bodenkunde. Systematik, Genese, Eigenschaften, Ökologie und Verbreitung von Böden. Stuttgart (Enke).

Hölzer, A. & A. Hölzer (1995): Zur Vegetationsgeschichte des Hornisgrinde-Gebiets im Nordschwarzwald – Pollen, Großreste und Geochemie. *Carolinea* 53: 199 – 228.

JAHN, R., HÄDRICH, F. & K. STAHR (1994): Bodenminerale in Raum und Zeit – Exkursionsführer zur Tagung der Kommission VII – Bodenmineralogie der Deutschen Bodenkundlichen Gesellschaft vom 5. - 8.10.1994 in Breisach am Kaiserstuhl. Hohenheimer Bodenkundliche Hefte 20 (II). Stuttgart (Selbstverlag).

LESER, H. (2009): Geomorphologie - 9. vollst. überarb. Auflage. Braunschweig (Westermann).

LOTTER, A.F. & H.J.B. BIRKS (1993): The impact of the Laacher See Tephra on terrestrial and aquatic ecosystems in the Black Forest, southern Germany. Journal of Quaternary Science 8 (3): 263 – 276.

MAHR, A. (1998): Lockerbraunerden und periglaziale Hangsedimente im Bayerischen Wald. Untersuchungen zu Paläoumwelt und Geomorphodynamik im Spätglazial und ihren Einfluß auf die Pedogenese. Regensburger Geographische Schriften 30. Regensburg (Selbstverlag).

MEYNEN, E., SCHMITHÜSEN, J., GELLERT, J., NEEF, E., MÜLLER-MINY, H. & J.H. SCHULTZE (1962): Handbuch der naturräumlichen Gliederung Deutschlands. Bad Godesberg (Selbstverlag).

METZ, R. (1977): Mineralogisch-landeskundliche Wanderungen im Nordschwarzwald - besonders in dessen alten Bergbaurevieren – 2. vollständig überarbeitete Auflage. Lahr (Schauenburg).

MEYER, B. & R. SAKR (1970a): Menge, Sitz und Verteilung der extrahierbaren Fe-, Al-, SiO_2- und Humusanteile und ihr Einfluss auf die Austausch-Eigenschaften von typischen sauren Lockerbraunerden. *Göttinger Bodenkundliche Berichte* 14: 49 - 83.

MEYER, B. & R. SAKR (1970b): Aggregate, Dispergierungs-Resistenz und Vorbehandlungs-Methoden zur Korngrößen-Analyse saurer allophanhaltiger Lockerbraunerden. *Göttinger Bodenkundliche Berichte* 14: 85 – 105.

PRIEHÄUSSER, G. (1966): Beiträge zur Frage der Entstehung von Lockerbraunerden im Bayerischen Wald auf pleistozäner Bodenunterlage. *Geologische Blätter für Nordost-Bayern und angrenzende Randgebiete* 16: 183 – 191.

RADKE, G.J. (1973): Landschaftsgeschichte und -ökologie des Nordschwarzwaldes. In: Alleweldt, G. [Hrsg.]: Hohenheimer Arbeiten, Reihe: Pflanzliche Produktion 68. Stuttgart (Ulmer).

REGELMANN, K. (1934): Erläuterungen zur Geologischen Spezialkarte von Württemberg, Blatt Enzklösterle (Nr. 78). Stuttgart (Riederer).

REGELMANN, K. (1935): Erläuterungen zur Geologischen Spezialkarte von Württemberg, Blatt Baiersbronn (Nr. 91) – 2. von Manfred Bräuhäuser überarb. Auflage. Stuttgart (Ernst Klett).

SAKR, R. & B. MEYER (1970): Mineral-Verwitterung und -umwandlung in typischen sauren Lockerbraunerden in einigen Mittelgebirgen Hessens. *Göttinger Bodenkundliche Berichte* 14: 1 – 47.

SCHAMMEL, C. (1991): Sedimentologische & biostratigraphische Untersuchungen an Sedimentkernen aus dem Profundal des Schurmsees (Nordschwarzwald). Freiburg (Selbstverlag).

SCHEFFER, F. & P. SCHACHTSCHABEL (2010): Lehrbuch der Bodenkunde - 16. Auflage. Heidelberg u.a. (Spektrum).

SCHMINCKE, H.U. (2000): Vulkanismus – 2. überarb. und erw. Auflage. Darmstadt (Wissenschaftliche Buchgesellschaft).

SCHÖNHALS, E. (1957a): Spätglaziale Ablagerungen in einigen Mittelgebirgen Hessens. *Eiszeitalter und Gegenwart* 8: 5 – 17.

SCHÖNHALS, E. (1957b): Eine äolische Ablagerung der Jüngeren Tundrenzeit im Habichtswald. *Notizblatt des Hessischen Landesamtes für Bodenforschung* 85: 380 – 386.

SEMMEL, A. (1993): Grundzüge der Bodengeographie - 3. überarb. Auflage. Stuttgart (Teubner).

SEMMEL, A. (1996): Geomorphologie der Bundesrepublik Deutschland – 5. überarb. und regional erw. Auflage. Erdkundliches Wissen 30. Stuttgart (Steiner).

STAHR, K. (1979): Die Bedeutung periglazialer Deckschichten für Bodenbildung und Standorteigenschaften im Südschwarzwald. Freiburger Bodenkundliche Abhandlungen 9. Freiburg (Eigenverlag).

STÖHR, W.T. (1963): Der Bims (Trachyttuff), seine Verlagerung, Verlehmung und Bodenbildung (Lockerbraunerden) im südwestlichen Rheinischen Schiefergebirge. *Notizblatt des Hessischen Landesamtes für Bodenforschung* 91: 318 – 337.

VÖLKEL, J. (1991): Bodentypen und -genese auf jungpleistozänen Deckschichten im Bayerischen Wald. *Mitteilungen der Deutschen Bodenkundlichen Gesellschaft* 66 (II): 877 – 880.

VÖLKEL, J. (1995): Periglaziale Deckschichten und Böden im Bayerischen Wald und seinen Randgebieten als geogene Grundlagen landschaftsökologischer Forschung im Bereich naturnaher Waldstandorte. Zeitschrift für Geomorphologie Supplement 96. Berlin & Stuttgart (Borntraeger).

WIECHMANN, H. (2000): Böden als Naturkörper – Podsole. In: Blume, H.P. et al. (1996): Handbuch der Bodenkunde - 9. ergänzende Lieferung. Landsberg (Ecomed).

WILMANNS, O. (2001): Exkursionsführer Schwarzwald – Eine Einführung in Landschaft und Vegetation. Stuttgart (Ulmer).

WINDHEUSER, H. (1977): Die Stellung des Laacher See Vulkanismus (Osteifel) im Quartär. Sonderveröffentlichungen des geologischen Instituts der Universität zu Köln 31. Köln (Selbstverlag).

7.2 Kartenverzeichnis

Landesvermessungsamt Baden-Württemberg (1963): Topographische Karte von Baden-Württemberg 1:50.000, Blatt L 7314 Baden-Baden. Stuttgart.

Bundesanstalt für Geowissenschaften und Rohstoffe (2002): Bodenübersichtskarte 1:200.000, Blatt CC 7910 Freiburg-Nord. Hannover.

Anhang 1: Untersuchungsräume und Lage der aufgenommenen Bodenprofile

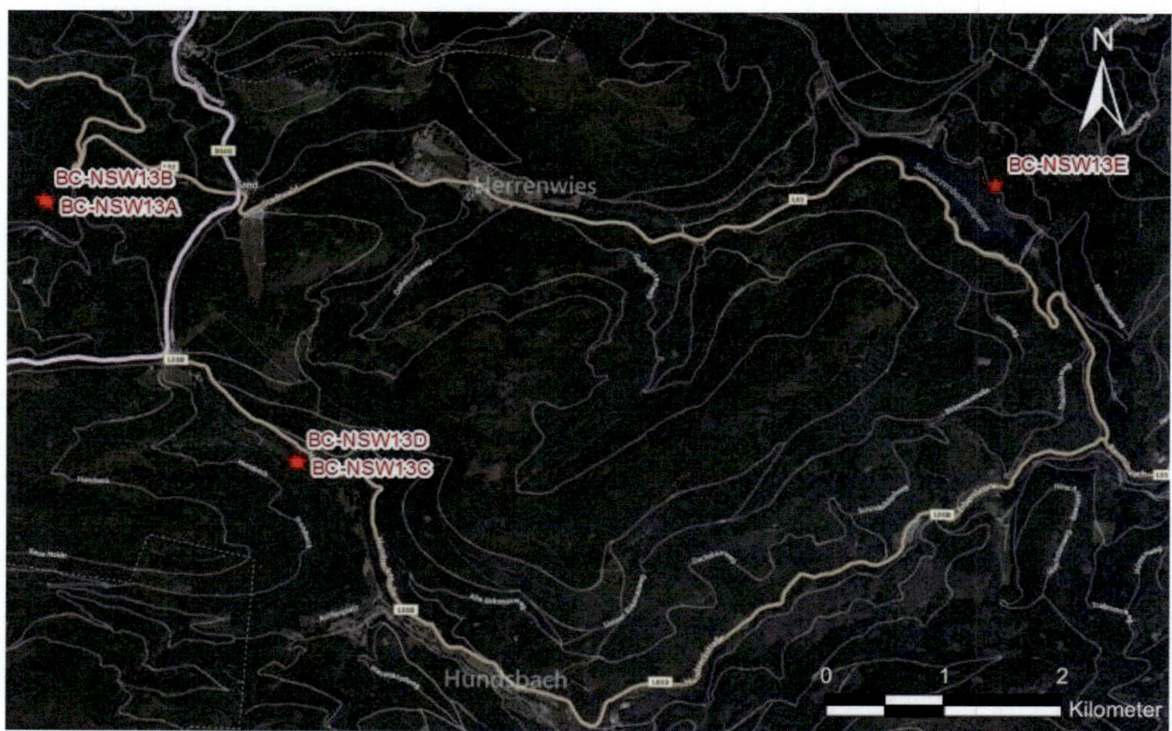

Anhang 1-1: Lageübersicht der Bodenprofile BC-NSW 13 A bis BC-NSW 13 E.

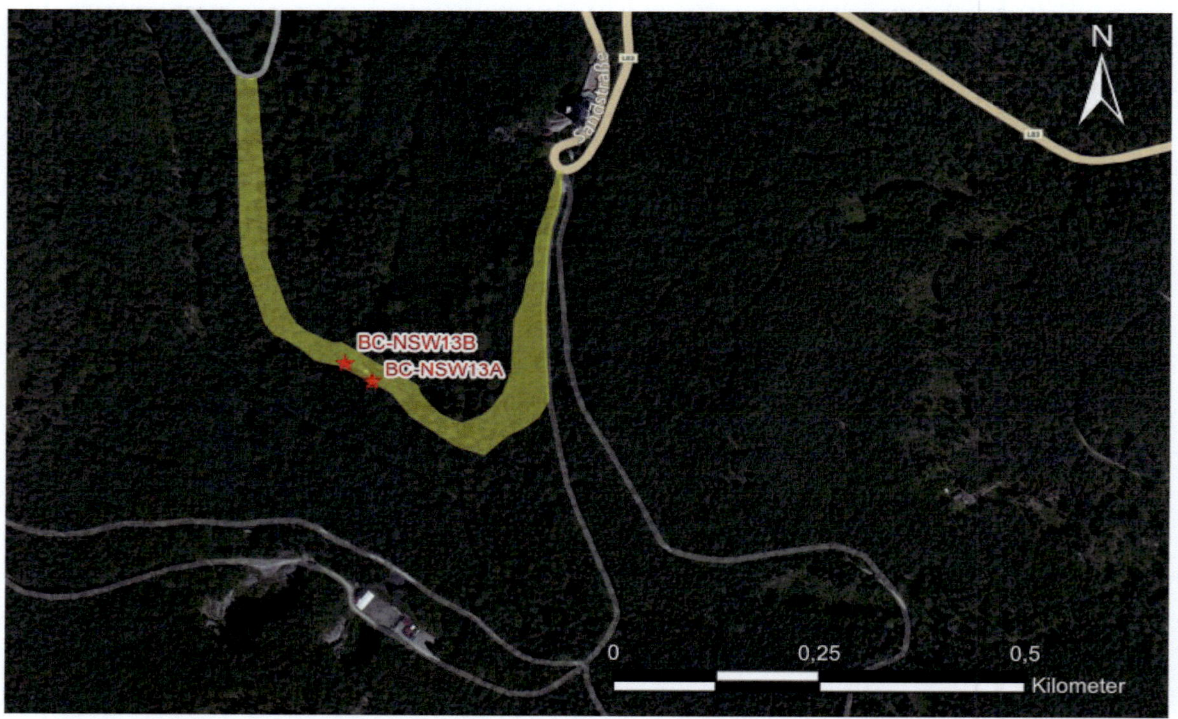

Anhang 1-2: Beprobungsareal 1 westlich von Sand mit Lockerbraunerdefundort BC-NSW 13 A.

Eigene Darstellung mit ArcGIS 10.1 auf Grundlage von Bing-Maps (2013).

Anhang 1-3: Beprobungsareal 2 südöstlich Hundseck mit Lockerbraunerdefundort BC-NSW 13 C. Eigene Darstellung mit ArcGIS 10.1 auf Grundlage von Bing-Maps (2013).

Anhang 1-4: Beprobungsareal 3 an der Schwarzenbachtalsperre mit Lockerbraunerdefundort BC-NSW 13 E. Eigene Darstellung mit ArcGIS 10.1 auf Grundlage von Bing-Maps (2013).

Anhang 2: Profilbeschreibungen gemäß verkürztem Formblatt der AG Boden 2005

Geländebefunde Profil BC – NSW 13 A

Zone 32 N: 442330 E, 5389650 N; 610 m über NHN

Gertelsbach-Wasserfälle, oberhalb der Gertelsbachhütte

Vegetation: Buche, Fichte, Eiche, Frauenhaarmoos

Geologie: Granit des Zweiglimmergranits, teilweise Zerblockung

Horizont	Mächtig-keit	Farbe	Bodenart	Skelett-gehalt	Gefüge	Durch-wurzelung	Humus-gehalt	Carbonat-gehalt
L	0,2	-	-	-	-	-	-	-
Of	0,5	-	-	-	-	-	-	-
Ah	-10	dbnsw	Sl4	Gr2	koh	Wg2/Wf2	h2	-
Bfv	-80	ocbn	Sl4	Gr2/X1	koh	Wg3/Wf3	h1	-
II ilCv	80 +	ochbn	Ls4	Gr5	koh	Wg1/Wf0	-	-

Humusform: F-Mull

Bodenform: Lockerbraunerde aus einer LH (Zweiglimmergranit, Löss & LST)
über einer LB (Zersatz des Zweiglimmergranit)

Geländebefunde Profil BC – NSW 13 B

Zone 32 N: 442320 E, 5389670 N; 600 m über NHN

Gertelsbach-Wasserfälle, unterhalb der Gertelsbachhütte

Vegetation: Buche, Fichte, Eiche

Geologie: Granit des Zweiglimmergranits, Zerblockung in weiterer Umgebung

Horizont	Mächtig-keit	Farbe	Bodenart	Skelett-gehalt	Gefüge	Durch-wurzelung	Humus-gehalt	Carbonat-gehalt
L	0,5	-	-	-	-	-	-	-
Of	1,0	-	-	-	-	-	-	-
Aeh	-7	grlisw	Ls4	Gr2	koh	Wg2/Wf2	h2	-
Bsv	-52	gebn	Ls4	Gr3	koh	Wg2/Wf1	h1	-
II ilCv	52 +	gehbn	Ls4	Gr5/X2	koh	Wg1/Wf0	-	-

Humusform: F-Mull

Bodenform: podsolige Braunerde aus einer LH (Zweiglimmergranit, Löss & LST)
über einer LB (Zersatz des Zweiglimmergranit)

Geländebefunde Profil BC – NSW 13 C

Zone 32 N: 444390 E, 5387290 N; 780 m über NHN

zwischen Großbach und L 80b bei Hundseck

Vegetation: Buche, Fichte, Tanne, Frauenhaarmoos, Waldschwingel

Geologie: Granit des Zweiglimmergranits, teilweise Zerblockung

Horizont	Mächtig-keit	Farbe	Bodenart	Skelett-gehalt	Gefüge	Durch-wurzelung	Humus-gehalt	Carbonat-gehalt
L	0,2	-	-	-	-	-	-	-
Of	0,5	-	-	-	-	-	-	-
Ah	-5	grlisw	Sl4	Gr2	koh	Wg1/Wf2	h2	-
Bfv	-40	hbnlioc	Ls3	Gr2	koh	Wg2/Wf2	h1	-
II ilCv	40 +	dbn	Ls3	Gr3/X1	koh	Wg0/Wf0	-	-

Humusform: F-Mull

Bodenform: Lockerbraunerde aus einer LH (Zweiglimmergranit, Sandstein des Buntsandsteins,
              Löss & LST) über einer LB (Zersatz des Zweiglimmergranits)

Geländebefunde Profil BC – NSW 13 D

Zone 32 N: 444380 E, 5387280 N; 605 m über NHN

zwischen Großbach und L 80b bei Hundseck

Vegetation: Buche, Fichte, Tanne, Sauerklee, Waldschwingel

Geologie: Granit des Zweiglimmergranits, Zerblockung in weiterer Umgebung

Horizont	Mächtig-keit	Farbe	Bodenart	Skelett-gehalt	Gefüge	Durch-wurzelung	Humus-gehalt	Carbonat-gehalt
L	0,2	-	-	-	-	-	-	-
Of	1	-	-	-	-	-	-	-
Oh	0,2	-	-	-	-	-	-	-
Aeh	-5	grlisw	Sl3	Gr3	koh	Wg2/Wf1	h2	-
Bsv	-40	vilibn	Sl3	Gr3	koh	Wg1/Wf1	h1	-
II ilCv	40 +	orlibn	Sl4	Gr2	koh	Wg0/Wf1	-	-

Humusform: mullartiger Moder

Bodenform: podsolige Braunerde aus einer LH (Zweiglimmergranit, Sandstein des
              Buntsandsteins, Löss & LST) über einer LB (Zersatz des Zweiglimmergranits)

Geländebefunde Profil BC – NSW 13 E

Zone 32N: 450450 E, 5389840 N; 690 m über NHN

hangoberhalb des Nordufers der Schwarzenbach-Talsperre

Vegetation: Buche, Fichte, Frauenhaarmoos, Waldschwingel

Geologie: Granit des Zweiglimmergranits

Horizont	Mächtig-keit	Farbe	Bodenart	Skelett-gehalt	Gefüge	Durch-wurzelung	Humus-gehalt	Carbonat-gehalt
L	0,5	-	-	-	-	-	-	-
Of	1,5	-	-	-	-	-	-	-
Oh	0,2	-	-	-	-	-	-	-
Ah	-15	vilibnsw	Sl4	Gr3/X1	koh	Wg2/Wf3	h2	-
Bfv	-65	oclibn	Ls4	Gr3	koh	Wg1/Wf2	h1	-
II ilCv	65 +	vilibngr	Ls4	Gr6	koh	Wg0/Wf0	-	-

Humusform: feinhumusartiger Moder

Bodenform: Lockerbraunerde aus einer LH (Zweiglimmergranit, Sandstein des Buntsandsteins, Löss & LST) über einer LB (Zersatz des Zweiglimmergranits, Sandstein des Buntsandsteins)

Anhang 3-1: Bodenphysikalische Laborergebnisse

Profil	BC-NSW 13 A		BC-NSW 13 B		BC-NSW 13 C		BC-NSW 13 D		BC-NSW 13 E	
Horizont	Ah	Bfv	Aeh	Bv	Ah	Bfv	Aeh	Bv	Ah	Bfv
Trockenrohgewicht [g]	237,31	186,47	225,08	286,56	203,65	233,42	242,94	279,71	230,26	234,22
Trockenrohdichte [g/cm³]	0,95	0,75	0,90	1,15	0,81	0,93	0,97	1,12	0,92	0,94
Porenvolumen [%]	55,36	69,42	60,83	52,88	65,60	61,84	59,23	54,37	60,45	60,30
Körngrößenanteil [%] gS	31,23	19,65	21,25	16,16	35,42	22,78	38,47	39,09	36,65	32,88
mS	13,71	15,03	10,67	11,67	18,01	17,01	24,28	25,03	19,54	19,19
fS	8,89	10,41	8,64	8,69	8,50	8,54	10,48	11,44	8,81	9,22
gU	14,40	18,48	16,71	20,59	7,38	9,08	6,60	5,81	6,05	5,81
mU	11,44	13,38	15,50	14,58	8,29	12,53	6,47	6,65	7,59	8,00
fU	6,92	6,88	7,14	6,49	5,70	9,08	4,36	3,87	5,41	5,16
T	13,41	16,18	20,08	21,81	16,70	20,98	9,34	8,10	15,96	19,74
Bodenart	Sl4	Sl4	Ls4	Ls4	Sl4	Ls3	Sl3	Sl3	Sl4	Ls4

grau = vermuteter Probenahmefehler

Anhang 3-2: Bodenchemische Laborergebnisse

Profil		BC-NSW 13 A		BC-NSW 13 B		BC-NSW 13 C		BC-NSW 13 D		BC-NSW 13 E	
Horizont		Ah	Bfv	Aeh	Bv	Ah	Bfv	Aeh	Bv	Ah	Bfv
pH		3,66	4,23	3,80	4,33	3,60	4,14	3,34	3,60	2,98	3,69
Corg		10,40	4,05	9,46	3,31	7,55	5,00	7,14	3,00	6,88	5,43
oxalatlösliche pedogene Oxide [%]	Fe_o	0,445	0,790	0,534	0,420	0,783	0,788	0,279	0,253	0,534	0,658
	Mn_o	0,005	0,011	0,034	0,038	0,001	0,006	0,001	0,001	0,001	0,002
	Al_o	0,321	0,619	0,469	0,642	0,573	1,272	0,137	0,138	0,172	0,396
dithionitlösliche pedogene Oxide [%]	Fe_d	0,769	1,265	0,987	0,990	1,292	1,463	0,611	0,495	0,932	1,151
	Mn_d	0,008	0,016	0,037	0,040	0,003	0,015	0,002	0,002	0,003	0,006
	Al_d	0,320	0,673	0,476	0,641	0,766	1,385	0,137	0,141	0,216	0,466
Gesamtgehalte pedogener Oxide [%]	Fe_{ges}	1,214	2,055	1,521	1,410	2,075	2,251	0,890	0,748	1,466	1,809
	Mn_{ges}	0,013	0,027	0,071	0,078	0,004	0,021	0,003	0,003	0,004	0,008
	Al_{ges}	0,641	1,292	0,945	1,283	1,339	2,657	0,274	0,279	0,388	0,862
Quotienten der pedogenen Oxide	Fe_o/Fe_d	0,579	0,625	0,541	0,424	0,606	0,539	0,457	0,511	0,573	0,572
	Mn_o/Mn_d	0,625	0,688	0,919	0,950	0,333	0,400	0,500	0,500	0,333	0,333
	Al_o/Al_d	1,003	0,920	0,985	1,002	0,748	0,918	1,000	0,979	0,796	0,850